SUSTAINABLE GARDENING

SUSTAINABLE GARDENING

Grow a "Greener" Low-Maintenance Landscape with Fewer Resources

Vincent A. Simeone

First Published in 2021 by Cool Springs Press, an imprint of The Quarto Group,
100 Cummings Center, Suite 265-D, Beverly, MA 01915, USA.
T (978) 282-9590 F (978) 283-2742 QuartoKnows.com

Cool Springs Press titles are also available at discount for retail, wholesale, promotional, and bulk purchase. For details, contact the Special Sales Manager by email at specialsales@quarto.com or by mail at The Quarto Group, Attn: Special Sales Manager, 100 Cummings Center, Suite 265-D, Beverly, MA 01915, USA.

25 24 23 22 21 1 2 3 4 5

ISBN: 978-0-7603-7036-0

Digital edition published in 2021

eISBN: 978-0-7603-7037-7

Some of the content in this book appeared in the previously published *Grow More with Less* by Vincent A. Simeone (Cool Springs Press 2013).

Library of Congress Cataloging-in-Publication Data available
Cover Images: Shutterstock
Page Layout: Sporto
Photography: Courtesy of author and Shutterstock

Printed in China

This book is dedicated to all of the talented and passionate green industry professionals who continue to devote their lives to the noble cause of protecting our natural world. So many of my fellow colleagues continue to fight the good fight by pouring their lives into making the world a better place. Now more than ever cultivating our gardens, growing our own food, and creating habitat for wildlife are vitally important to our future. We must never forget that our environment is fragile and easily compromised by our actions and that humans are only a small part of a much bigger picture. Without question, the best part of my career as a horticulturist is working alongside and learning from so many talented and exceptional people.

Most important, this book is dedicated to the memories of Fritz Schaefer and Pat Voges, two legends in the green industry who left an indelible mark on so many and who not only cultivated plants but also were masters at cultivating relationships for the greater good.

CONTENTS

INTRODUCTION

The term *sustainability* is a catchy word that has garnered a lot of attention the past few years. But unlike many catchphrases, sustainability has substance, credibility, and the potential to change life as we know it. For those who accept sustainability as a way of life, it will undoubtedly pay long-term dividends. Sustainability is a complex issue that affects every facet of our lives. It requires an investment of time and resources and yields positive results over many years. This commitment of time, hard work, and patience is a worthwhile effort that reduces the impact we human beings have on our environment. Like many worthwhile endeavors, sustainability offers us a chance to lead a responsible, healthy life and provide an example for others to learn from.

Environmental destruction is one of several reasons why sustainability has become such an important issue. Many notable scientists and environmental advocates believe that our impact on the world around us has a direct and negative effect on global climate change. Human nature is to advance both intellectually and physically, and that all too often means more building, more consumption, and more resources needed. This progress comes with a price, though, usually at the cost of the environment. In the past, while we built roadway systems, housing developments, shopping malls, and resorts, the long-term impact on the environment was often not considered or anticipated. But that is slowly changing, and everything from large developments to residential settings has the ability to become more sustainable.

An example of a well-designed landscape with multiple seasons of interest

Replacing resource-intensive turf grass with low-maintenance flowering plants is a simple way to help the environment.

For the home gardener, the sustainability movement has taken giant steps forward over the past decade. The concept of becoming more environmentally sensitive has gained momentum to the point where it now seems to be an unavoidable, irresistible force. It's no revelation that many of us have been unknowingly practicing sustainable living for many years without even realizing it. The difference is that today, sustainability isn't just an option anymore; it truly is a necessary way of life. The days of casually practicing sustainable living when it is convenient are in the past. The time has come to embrace sustainability as an everyday part of life.

Sustainability is a relative and subjective term that can mean different things to different people. Sustainability will be influenced by many factors, including economics, the size and complexity of your property, climate, and available resources. It is difficult to put a specific definition on the term with the idea that it will apply to all, or even most, situations that we encounter daily. Because we live in such a dynamic, changing environment with an incredibly diverse population, the concept of sustainability will need to change and adapt, along with us as a society. In order to successfully nurture a more sustainable environment, we must understand what it takes to create a lifestyle that supports this way of living.

DEFINING SUSTAINABILITY

By definition, sustainability is the capacity to endure. It is the belief that living systems can persevere and adapt no matter what the circumstances. For humans, sustainability is the long-term care and maintenance of the world around us—which has environmental, economic, and social implications—and encompasses the concept of stewardship and the responsible management of natural resources. Most important, the sustainability movement must include the idea that what we do today should not negatively affect future generations. In the context of nature, sustainability relates to how biological systems remain diverse and productive over time, a necessary element for the long-term well-being of humans and other organisms.

With threatened ecosystems, limited natural resources, and a changing environment, developing more sustainable ways to live has become a necessity. For example, ecotourism has become very popular and economically viable for communities around the world, but the human interaction and impact in these environmentally sensitive areas requires careful oversight. If an old-growth forest or a diverse barrier reef is being negatively impacted by human activity, there are ways to manage these issues. These management techniques include limited or seasonal use or alternate use that reduces impact. In the case of hunting and fishing, the legal limits put on the numbers of fish or deer that can be harvested during a given season is an excellent example of sustainable practices. Without these guidelines, animal populations would be greatly reduced or become extinct. In the grand scheme of things, sustainability evokes the reality that humans are a smaller part of a bigger picture and we must do our part to ensure the viability of future generations. There has long been a debate about where we humans fit in the hierarchy of the environmental system. The reality today is that what we do ultimately affects the world around us and we must do our part to protect the symbiotic relations we have with the Earth.

As consumers and users of the land, humans have both positive and negative impacts to the environment. There is some debate on whether recent climate change has been caused by human activity, but there is little debate that it is occurring. Many experts feel that global climate change is causing severe weather patterns, droughts and floods, and variations in temperature patterns that are impacting where plants grow. Whether this is a short-term or long-term trend remains to be seen. Obviously, our hope is that the weather patterns return to "normal" and that what we are currently experiencing is just a minor speed bump in terms of long-range weather trends.

One common way of measuring the effect that humans have on their environment is known as a carbon footprint. A carbon footprint offers us the opportunity to quantify the impact that we have with the resources we use every day. A carbon footprint is measured as the greenhouse gas (GHG) emissions that result from an organization, activity, product, or people. GHGs can be emitted through transportation, clearing of land, and the production and consumption of food, fuels, manufactured goods, materials, roads, buildings, and so forth. These activities cause the emission of harmful amounts of carbon dioxide or other GHGs. The negative effects of these damaging gases can even be generated from household power equipment such as lawnmowers, leaf blowers, weed trimmers, or any equipment that uses fossil fuels. Of course, cars, the burning of heating oil, and factories that create harmful pollutants to manufacture products can all have huge impacts on the environment. As we know from popular terms such as global warming or the greenhouse effect, these emissions can contribute to climate change.

We now know that the impact of human consumption and activity affects our entire ecosystem. This realization has most certainly accelerated the need to look for ways to be more responsible.

Even small actions, such as growing some of your own food and planning your garden with eco-friendly techniques in mind, can add up to a big reduction in your carbon footprint.

SUSTAINABILITY FOR THE HOME GARDENER

On a smaller and more comprehensible scale, the principles of sustainability can *definitely* be applied to your home landscape. What we grow, how we grow, and how we manage our gardens are important factors to consider when it comes to encouraging sustainability. For far too long, sustainability concepts have not been incorporated into the planning and development of residential or commercial landscapes. *You*, as the designer of a new or completely renovated landscape, have the opportunity to start fresh and incorporate the innovations that sustainable techniques have to offer. Even if some landscapes receive only a partial makeover, sustainable practices can be readily incorporated, resulting in a better garden. In many cases, gardeners are attempting to enhance the existing conditions of a well-established landscape. If this is the case for you, trying to improve or change years of accumulated or inherited mishaps and poor gardening practices can be a challenging and often frustrating task. Don't give up. The key is to make these improvements part of an overall strategic plan that is manageable and prioritized. Superior and innovative design techniques work hand in hand with proper sustainability initiatives.

In addition to the design of the landscape, sustainability is an integral part of many other aspects of gardening, including plant health care, recycling, composting, organic gardening, energy efficiency, alternative fuels, fuel-efficient equipment, and water conservation.

Gardening is extremely popular among a large portion of the worldwide population on both professional and hobbyist levels. While gardening originated in Europe and Asia centuries ago, it continues to evolve along with new technologies and research. But with an ever-changing climate that often includes extreme and unpredictable weather conditions, finite and often strained resources, and increased pressures such as new pest problems, gardening has become increasingly challenging. So it's no surprise that sustainable gardening has emerged as an innovative, efficient, and responsible way to grow plants and manage landscapes. Until recently, the concept of living a more sustainable lifestyle in the garden had mostly been an afterthought in many households. Thankfully, that is rapidly changing—first in the professional horticultural and landscape design industry and now in the home garden.

Within a garden setting, sustainability directly relates to the proper management of soil, composting to enhance soil properties, responsible water usage, sound maintenance practices for plants, managing invasive species, and most important, proper plant selection. Putting the right plant in the right place is not only important, it is essential to garden success. Sound landscape design to support a lower-maintenance, healthy, and functional landscape is equally important.

Cultivated landscapes can be layered with different levels of vegetation, just like the forest.

One of the most important components of sustainability is the development and selection of plants that require less care and resources. These are plants selected for their genetic superiority, whether it's in the form of improved ornamental qualities, pest resistance, drought tolerance, cold tolerance, or other traits. New varieties of landscape plants and agricultural crops are regularly being developed to accommodate the needs of a growing and more demanding population of gardeners. Today, all types of plants are being developed for the garden from new species and varieties of trees, shrubs, annuals, perennials, and even agricultural crops that can also be used as ornamentals in the garden. Plant selection and diversity are two of the most important components in an effective sustainability program to keep landscapes healthy and thriving.

These and other sound gardening techniques are vital to the long-term success of our landscapes, and they are the primary focus of this book. Sustainable gardening reaches far beyond traditional horticultural practices of gardening for aesthetics. In fact, it is more about the function of landscape plants these days than just how pretty the plants are. While having a colorful and aesthetically pleasing garden is still a priority, reducing the resources needed to maintain that beauty is the end goal. If sound, responsible, and well-planned gardening techniques are employed, utilizing sustainable practices, dedicated gardeners can have their cake and eat it, too.

THE KEY TO SUSTAINABLE GARDENING

Sustainable gardening is influenced by many factors, including size of the garden, budget, climate, soil conditions, and local regulatory restrictions. A large private estate or public municipality will have more resources available than a small home garden. However, successful sustainable gardening isn't dictated only by economics. You can accomplish your desired goals by following the consistent, sound gardening principles outlined in this book. They apply to large, commercial sites as well as smaller, residential ones. Essentially, a good, well-thought-out plan is the blueprint to success regardless of the size or scope of a project. I have visited many gardens throughout the world, and some of the most impressive ones have been residential landscapes with a clear focus and goal. By creating a solid, functional design and good foundation, any landscape is bound to succeed.

LESS IS MORE: THE IMPORTANCE OF THE SUSTAINABILITY MOVEMENT

You may wonder why sustainable gardening is so important. It's important for the same reason living a more responsible, sustainable lifestyle is. Less is more, and gardeners have to find creative ways to do more with less. Today, natural resources are more limited than ever and the competing demand for them is at an all-time high. With rising costs for fossil fuel and other forms of energy, foods, and products, efficiency is essential. The impact humans have made on the environment over the last century proves that conservation, preservation, reducing water needs, and using fewer resources is essential to our future. With a growing population and increased demands on our environment, more attention needs to be paid to protecting water supplies, reducing chemical use and pollution, managing invasive species, and nurturing animal and plant life.

This well-balanced landscape is designed in layers to offer a succession of bloom as well as a multi-tiered arrangement with larger plants in the back and smaller plants in the front.

Generally speaking, sustainability is the capacity to endure over the long term. This low-maintenance environment will ultimately support beneficial insects, birds, and other important wildlife, and also reduce the impact of invasive plants and the reliance on chemicals. It also encourages a well-balanced landscape through the use of organic maintenance measures. While the short-term benefits of chemicals and traditional gardening practices are tempting, more sustainable methods offer greater options and alternatives. Sustainable gardening is, without a doubt, a "marathon" that requires endurance and patience rather than the shorter-term "sprint" or quick fix that has been the main theme within the home garden for the past few decades.

MORE ON CLIMATE CHANGE

One of the most important issues influencing the world today is climate change. Many experts feel there is consistent evidence to suggest that the Earth is getting warmer and climate change is a reality we must face now. But more than just a warming trend, there is a bigger picture in terms of what climate change means to us. Climate change is thought to affect overall weather patterns, causing extreme weather conditions, variations in temperature (both hot and cold), drought, excessive rain, and flooding. In the gardening world, climate change is believed to be affecting plant growth and blooming times, along with pest ranges and life cycles.

While it is impossible to predict what the future holds, many environmental groups and worldwide meteorological organizations are closely monitoring this issue and the future impact it will have on our environment. This is why the sustainability movement is so important right now. Many scientists believe that reducing the human impact to the environment will decrease these effects or, at the very least, not make them any worse in the future. Climate change adds a sense of urgency to becoming more sustainable, especially from a landscape perspective. Practicing smart horticulture and sound landscape design practices can play major roles in improving the community we live in. Protecting our natural resources, wildlife habitats, and plant life—one garden at a time—will undoubtedly help the cause. It's especially important, since large corporations, farmers, golf course managers, nurseries, and just about everyone around us has already started some sort of sustainability program.

There's evidence that climate change could be affecting some less desirable plants, such as this poison ivy, by encouraging more active growth.

LOW MAINTENANCE IS A GOAL

Low-impact gardening methods lead to a more sustainable landscape for many generations to come.

Low-maintenance gardening is one of the main benefits of developing a sustainable landscape and vice versa. The two are integrated, but this takes an investment of time and patience, at least initially. While many gardeners enjoy working in their gardens, reducing maintenance is worthwhile for several reasons. First, by reducing the overall maintenance of the garden, gardeners can focus their time and efforts on future planning, enhancement, and expansion of the garden, as well as the acquisition of new plants and garden features. Second, a low-maintenance garden requires fewer resources and will ultimately cost less to care for. If these landscapes are maintained using a more natural, organic approach in a holistic, comprehensive manner, the long-term benefits include saving resources, reducing health risks, and ultimately reducing intense maintenance and day-to-day care. How much time is needed to accomplish this really depends on the type, size, and complexity of the garden and the environmental conditions in it. But in general, even within a few years, great strides can be made to reduce garden maintenance if sustainable practices are implemented.

Some examples of using sustainable practices to reduce maintenance are proper mulching and composting to enhance soil biology, creating habitat for beneficial insects, and proper siting and selection of superior plant species. A few good examples of low-maintenance practices that will lead to a more sustainable landscape include raising mowing heights of turf, using a mulching mower, and designing the garden with smooth curving lines that are easy to mow along. Throughout this book, you'll learn more about these practices and be introduced to many more.

THE GARDEN AND BEYOND

So, what specific things can you do as a homeowner to practice more sustainable living? Using alternative fuel vehicles that run on renewable energy is quickly becoming a popular trend that is an efficient, reliable, and eco-friendly way of traveling. New vehicles can be powered by electricity and recharged, or powered by natural gas. New hybrid vehicles that run on a combination of gasoline and electricity have become quite popular. Their gas mileage is exceptionally good. New technology has resulted in electric or battery-operated power tools and gardening tools that reduce pollution and are easier to use. This includes mowers, blowers, weed trimmers, hedge trimmers, pole saws, and many other popular garden tools. All of these new innovations support the idea that sustainable living is much more within our grasp now than it was 20 years ago. New technology continues to be the driving force that guides us to a greener future. Need a few more ideas?

- **Compost.** Besides making free organic fertilizer, composting your household kitchen and yard waste reduces the amount of garbage that ends up in the landfill and reduces GHG emissions.
- **Recycle.** Recycling reduces energy consumption, because it saves on manufacturing new products and disposing of old ones.
- **Buy recycled.** Using products that were made from recycled products, such as composite lumber or recycled plastic used to make decking, benches, and garden pots, will put less of a burden on natural resources such as wood.

EMBRACING SUSTAINABILITY

There are many reasons and tangible benefits to going green in your home and garden. These benefits can include increasing energy efficiency and cost savings, increasing property value, and reducing health risks. In business, green products also help foster healthy competition, which helps our economy. In a garden setting, going green can translate into the reduction of exposure to harmful chemicals and materials and an overall healthier, thriving garden.

To make sustainability work for you, you must embrace the notion that human beings have a great effect on their environment and accept the fact that you are a part of a much larger equation. The idea that humans are at the top of the food chain and should be allowed to consume what we need without taking responsibility is irresponsible and shortsighted. It is important to understand that we need to respect and share our natural world and its resources with a vision toward protecting future generations. It is equally important to understand that we need animals, plants, clean air and water, a reliable food source, and healthy soil to survive. The environment is not ours to use and sometimes misuse as we see fit. The more products and materials that can be reused rather than wasted, the cleaner our environment will be. Sustainability gives us the opportunity to protect our natural resources while creating a healthier, more productive life for everyone. If there was ever a mantra that typifies what sustainability is all about, it is this: Rethink, reuse, and reduce.

Wise use of resources, recycling, and repurposing are some of the key elements of sustainable gardening.

CHAPTER 1

ECO-FRIENDLY STRATEGIC PLANNING FOR YOUR GARDEN

Setting Up a Blueprint for the Future

The term *eco-friendly* is used quite often, but do you truly know the meaning and intent of the term? Many people don't. *Eco-friendly* is an abbreviated version of *ecologically friendly* or *environmentally friendly* and basically means the activities, products, practices, and policies that do the least amount of harm to the environment. Companies and organizations have used this term regularly as a marketing tool to make their products more attractive to consumers. In the garden realm, *eco-friendly* can have very specific and direct significance. For example, compost and processed organic products are marketed as eco-friendly products because they are derived from plant and animal matter. It seems only natural that this recycling process of returning to the earth what came from it is the right thing to do. Eco-friendly products can include a wide range of recyclables, such as fertilizer, paper, and cardboard; recycled plastic and rubber products; recycled wood products; and more. Eco-friendly activities can range from using greywater (recycling wastewater as a source of nonpotable water to wash clothes or water the lawn) to riding a bike to school or the mall rather than using a car. Using energy conservation initiatives and equipment that reduces harmful emissions is not only eco-friendly but is economical as well.

Eco-friendly strategic planning takes the term *eco-friendly* to a whole new level and validates why sustainability is so meaningful. There is no question that in order to be successful at creating a sustainable garden, you have to do a little research, be prepared, and have a plan in place before starting. A gardener without a plan is destined to waste time and money trying to piece together various sustainability practices with no clear sense of direction. Planning gives you the opportunity to map out a schedule, prioritize, and list the goals that will undoubtedly make your life easier in the long run.

PLANNING FOR THE LONG TERM

As I mentioned in the book's introduction, long-term planning is a key element in developing a sustainable garden. Proper planning and design will enable your home landscape to survive the stress of an ever-changing, challenging world that includes extreme environmental conditions, increased pressure from pests, and other environmental challenges. If we have learned anything from the history of the cultivated landscape, it is that the landscape is a dynamic entity. The late part of the nineteenth century until the middle part of the twentieth century brought ingenuity and the realization that open space and the interaction between the landscape and humans is critical. Back then, landscapes were built with long-term goals and great aspirations of making a difference. But slowly our public and private landscapes have become shorter-term propositions with little regard for long-term viability. The ideals and life lessons that worked in our favorite community gardens can be applied to our own residential landscapes. As home gardeners, we love the open space of our own backyards and that connection with nature. This open space does not have to be a large, expansive lawn but can be something more long term, such as a diverse landscape planting that you can enjoy growing old with. It is time to start reinventing our gardens and infusing in them the same passion and long-term planning that previous generations felt so committed to.

Today, in the context of new plantings, many home gardeners focus on how their landscape looks *right now*. It's all about "How do I screen the neighbor's fence?" or "How fast will my perennial garden grow so I can have instant color in the garden?" But that is not how sustainability in the garden works. We should not be focused on the short-term benefits of the garden as much as what it will offer and look like years down the road. When you're designing your garden—and *before* you add plants—you've got to do some research to fully understand how the plants you are considering grow and what their ultimate size will be. Many times I have seen gardeners plant whatever they want wherever they want, with the idea that if the plant gets too big or becomes a problem, it can just be pruned or transplanted sometime in the future. A better approach would be to provide your plants with the space and growing environment that will maximize plant growth and function in the landscape. Plants that have been put in the right situation have a much better chance of thriving and requiring little care while providing big dividends.

When planting new trees and shrubs, be sure to properly space them to keep them from being crowded when they're mature.

The long-term planning of the landscape is especially important in planting trees, shrubs, and evergreens. These plants typically live for a long time compared to other plants, such as herbaceous and tropical plants. More often than not, these woody plants are not given adequate room to grow, outgrowing their allocated space way before they reach maturity. I have witnessed many an evergreen grouping shoehorned into an area too small to accommodate it, with spacing of only a few feet between plants. In five years' time, these plants will compete with one another and shade each other instead of complement the plants in the garden.

THE ONE-HALF RULE

One basic rule that I recommend, whether you're planting a shrub, shade tree, or evergreen, is the One-Half Rule. Whatever plant you select for your yard, find out how large it will ultimately grow. This can be done by looking at the nursery label or asking a sales associate at the garden center or nursery you are purchasing from. Assume the plant you are purchasing will grow *at least* half that size in your lifetime. This will help you determine how far the plant should be planted from the house, fence, garage, or other plants. For example, if you choose to plant a row of variety of Hinoki Falsecypress (*Chamaecyparis obtusa*), and the plant tags state they will grow 30 feet (9.1 m) tall and 15 feet (4.6 m) wide at maturity, then at the very least, you should assume that each plant would grow 15 feet (4.6 m) tall and 7.5 feet (2.3 m) wide in your lifetime. That means that you should plant these falsecypresses farther apart than just a few feet—in this case, more than 5 feet (1.5 m) apart. Yes, it will take longer for your plants to touch or fill in, but ample space will allow them to reach their ultimate size unobstructed by an overcrowded garden.

Obviously the One-Half Rule depends on the specific growth of your plants, your climate, your soil, and many other factors—but you get the idea. This rule may seem a bit extreme, but many modern landscapes become overgrown with poorly shaped plants because we want an instant landscape. This rule supports the idea of a more open, airy landscape with plenty of air circulation and light, which create better growing conditions.

Proper spacing, siting, plant selection, and understanding of the specific growing conditions for the plants in your yard are all important factors that must be considered when planning the garden. Plants that are given the proper growing environment and spacing will be healthier, require less maintenance, and offer the most function and ornamental qualities possible. The more tinkering you do to control the natural environment that plants live in, the more you create an artificial, unsuitable environment for your favorite garden plants. Planting for the long term ensures that the landscape will be able to survive, even through adverse conditions, for a long time.

CONDUCT A SITE ANALYSIS

A landscape site analysis doesn't have to be fancy. The goal is to record what you see and what changes you'd like to implement and build a plan from there.

The concept of site analysis is derived from a series of steps in urban planning. It involves research, analysis, and synthesizing findings to develop a plan. This formal and often complex process is used widely in architecture, landscape architecture, and engineering. However, within the context of garden sustainability, this can be a much simpler process. (The principles are still the same.)

The goal of a site analysis is to evaluate your garden and record your observations. This is critical before designing the garden and implementing that design. Whether you are planting or altering part or all of the landscape, a site analysis gives you the opportunity to fix past failures or issues. Without a site analysis, gardeners will likely repeat the same mistakes over and over because they've given too little time to the planning process. A sound, well-thought-out, organized strategic plan will lay the foundation upon which your sustainable garden will be built, making the most of your efforts. How many times have we purchased plants, pottery, garden ornaments, furniture, and so forth at a local store and just placed them where we had room? A thorough site analysis will help you determine where things should go in the garden—sort of like arranging a room in your house.

In addition, a site analysis gives you an opportunity to evaluate existing conditions in your garden. This is when you should be evaluating soil type and drainage capacity, light conditions, exposure and wind issues, erosion problems, and so forth. Site analysis will also enable you to reconfigure your garden if something is not working well for you. Perhaps your patio is in the wrong location, or you would prefer to have more shade on the western part of the house in the afternoon and need an evergreen screen.

When performing your site analysis, rarely are you dealing with a blank page that can be easily manipulated. More often you will be evaluating a site within an established landscape. If you move into a newly built home and you have a fairly blank slate with a lawn and a few new trees, it does make things a bit simpler. But if you own or inherit a mature landscape, you will have to decide how to work around existing features such as large trees, shrub borders, walkways, patios, and so on. This is all the more reason why a site analysis is so important in the process of creating a sustainable landscape. It gives you the option to erase past blunders and enhance the functionality, aesthetic qualities, and long-term sustainability of your garden.

Step 1: Draw a Simple Survey

Before you begin your site analysis, you will need a few materials to complete the survey of your property.

- Clipboard and notepad to document your findings
- Deed map (This is a simple map that will show property lines, outline of the house, and other important information.)
- Measuring tape and helper to measure bed dimensions, widths of walkways, placement of trees, and other important information
- Digital camera or smartphone to take "before" pictures, which will help you determine what you like and dislike about your garden

If a deed map of your property is not available, simply get some drawing paper and sketch out the basic areas of your property: house, driveway, walkways, any other significant landscape features, and so forth. If you are reconfiguring the garden, you would only want to draw in any existing features that will remain and then draw in any new features—plantings and such—that you want to obtain.

Step 2: Take a Plant Inventory

Other important information that you should be recording is a plant inventory. This is especially true if you plan to move or redesign a particular area of the garden. This inventory should include plants you want to salvage as well as ones you want to remove. All of this information should be noted on your sketch. A simple chart can be created to easily organize and record inventory of your garden.

By recording this valuable information, you will learn a lot about the plants that are growing in your garden. For example, if your hydrangeas are performing poorly in the area where you have them currently, it may be wise to look closer at the conditions they are growing in and move them to a more suitable area of the garden. Or if the Japanese maple is getting too close to the house and is in poor condition anyway, maybe it is time to replace it with something more suitable.

Make a list of the plants already on your property, noting which ones you want to save and which should be removed.

Step 3: Put Everything in Its Place

Once you have documented existing conditions of the site, you can determine where specific plants will grow best, where the lawn should go, where the compost bins would work best, and so on. Obviously you would want to site the vegetable garden or the lawn in a sunny location with good drainage and place shade-loving perennials and shrubs along the perimeter of the garden in more protected areas. Since sustainability is doing (and growing) more with less, maximizing the space that you have and finding the right place for plants, hardscape features, recreational areas, and even just an area to sit and enjoy your handiwork are important. Putting everything in its proper place will give the garden organization and proper perspective.

Other information to consider when doing a site analysis:

- Pay close attention to where underground utilities such as water, electric, telephone, and natural gas are. Call to have a utility markout done by a trained professional before doing any digging.
- Over several days, weeks, or months, observe how sunlight hits the garden and what areas are in shade, partial shade, or sun most of the day. Taking a photograph in the exact same spot each time you do your observation will give you an excellent perspective on how sunlight illuminates your garden over a period of time.
- Check for overhead wires, tree branches, fences, and other obstacles that may interfere with the canopies of new trees or shrubs.
- Take soil samples from various parts of the garden to determine soil types, fertility, and drainage. This will help you determine which plants—flowers, shrubs, trees, vegetables, or lawn—are best suited for a specific location.

A proper site analysis is the first step in creating a garden that will work best for you. Without one, you are walking around blindfolded and not seeing your garden for what it really is. Knowing what you have, what you need, and how to obtain it should be priority number one. You will not be able to truly maximize the potential of the garden without this critical first step.

SIZE REALLY DOESN'T MATTER

As you have already learned, sustainability has no real boundaries and can be applied to the smallest postage stamp–sized garden as well as the large, expansive landscape. Sustainability can be valuable no matter what size, shape, or type of garden you prefer. What is more important than size is the know-how to make the most of the resources you have available to you in your specific situation. Not all gardens are created equal, with each individual at a different skill level and expertise. The common thread we all have in our fascination with gardening is the evolution of the landscape that always entices us to want more.

What is also important is not only how your garden is arranged but also how you maintain it. Overusing chemicals, misusing water, neglecting soil health, and planting and pruning improperly can all contribute to a poorly performing garden. So, once you have set up your garden the way you like, and once it is designed with sustainability in mind, it is time to implement those ideals that make sustainability so effective. Keep working at improving your garden and do not complicate what should be a rather simple and sensible approach to garden maintenance.

Whether your garden is small or large, in the ground or in containers, designing with sustainability in mind is essential from the start.

RIGHT PLANT, RIGHT PLACE

When choosing plants for your garden, carefully match the available growing conditions with plants that will perform best in those conditions. Here, shade-loving hostas and anemones are tucked under the tree canopy.

It sounds simple enough, but the idea of using the right plant in the right place has not always been as commonly utilized as one might think. This catchphrase embodies what sustainability and sound gardening are all about. I first heard the phrase over 30 years ago from two legendary nurserymen: Fritz Schaefer and Jim Cross. Since then, I have tried to practice that ideal whenever I select or recommend plants for a particular landscape.

Proper plant selection is one of the most important aspects of the sustainability movement. It requires patience, passion, and the ability to research superior species and varieties of plants and their optimal growing conditions. I have met gardeners from all over the world and from all walks of life, and they all have one thing in common: All of them have an insatiable appetite for new, exciting, and superior plants and for growing them to the best of their ability. This phenomenon does not just happen by osmosis but exists because each gardener has a spark that gets ignited every time he or she sets foot in the garden.

Gardening should be fun, satisfying, and educational; and choosing the right location for new plants can be both challenging and exhilarating. While it is always good to experiment and try different plants in different landscape situations, it is also not wise to place plants with reckless abandon. Some thought needs to go into why you purchased the plant in the first place, where it will grow best, and how it will be maintained. Many times gardeners make the mistake of selecting plants based solely on their aesthetic qualities, with the main consideration being "Where will the plant look best?" The answer to that question is that the plant will look best in the growing conditions it is most suited to grow in. A happy plant will be a productive and attractive plant. Placing a shade-loving plant that prefers moist soil in a hot, dry area of the garden because that's where it is most visible or where there is adequate space is a foolish proposition. Putting plants in areas of the garden where they like to grow will create a lower-maintenance, more functional, and ornamental landscape feature that will offer long-term benefits. Chapter 2 takes a deep dive into the concept of "right plant, right place" and introduces some of my favorite plants for a sustainable garden.

EXOTICS VERSUS NATIVES

It is important to mention that the introduction of exotic plant species and their impact on native ecosystems is a hot topic of debate. There is no doubt that many introduced insects, animals, and plants have caused serious harm to our natural environment over the years; but it is also important to point out that a plant that is an exotic species from another country is not automatically a threat to our ecosystem. In a natural setting, I believe that by all means we should be planting species that are indigenous to that particular area. But in a cultivated garden setting, noninvasive exotic species, natives, and cultivated varieties of natives are perfectly acceptable. Remember, a diversely planted, well-designed garden is a healthy garden. If you are unsure if a particular exotic species is invasive in your area, contact your local nursery or environmental protection agency or agricultural service.

ESTABLISH PRIORITIES

As you now know, creating a sustainable garden requires planning, organization, and the ability to follow through and implement your plan. The planning process is critical to success because it sets the foundation upon which you can build the garden. When planning to develop a sustainable garden, it is wise to start simple and gradually add more complex components as it develops. Gardeners should not fall into the trap of biting off more than they can chew, especially when resources are limited. Once you have a sound garden site analysis, design, and plan, you can then implement goals in priority order. It's okay to think big, but it's prudent to tackle one priority at a time instead of trying to do everything at once.

My recommendation is to design and implement any infrastructure renovations, including walkways, patios, irrigation installation, home improvements, and so on before moving on to landscaping. Once this infrastructure is in place, you can then site trees, shrubs, hedges, perennial borders, and so on.

When you are designing and planning part or all of your garden, here are a few additional guidelines I recommend you follow.

1. If you are totally renovating your garden, or if you have a blank slate, it is important to lay out the garden first. Sketching out where things should go based on the information taken from your site analysis will help you formulate your plan.

2. A well-designed garden should be functional and well balanced, with different components integrated seamlessly. For example, the lawn area should transition into flower beds, shrub borders, and ornamental plantings, while work areas such as vegetable gardens and cutting gardens might be better placed in the rear or along the side of the house near the compost pile and other related activities.

3. It is wise to organize your garden so that related garden features are grouped together. More often than not, gardeners find themselves altering only a portion of the garden; but it is important to ensure that whatever elements you are adding are properly placed and integrated well with their surroundings. If you choose to have an open lawn area in the front of the house, with a view from the kitchen window, you would not want to place a berm of evergreens in the middle of the front yard, as it would look awkward and would eventually hide the view that you are trying to maintain.

Whether your sustainable garden is large or small, use design guidelines and a site analysis to plan your garden long before you start to dig.

SET UP A GARDEN SCHEDULE OR PLANNER

Once you have an idea of how you want your garden to look and what features it should offer, it is time to set up a schedule of activities that will help you organize what happens in the garden and when. This garden planner will help guide you in future activities and remind you of past activities in the garden. For example, a garden schedule or planner can be used to document when you planted your first crop of warm-season vegetables in the spring or remind you when you need to fertilize your trees or herbaceous plantings. One of the advantages of doing this is that it offers a guide on what to

Sample Garden Maintenance Schedule

Type of Activity	Early Winter	Midwinter	Late Winter	Early Spring	Mid-Spring
PLANTING					
Container grown				X	X
Balled and burlapped				X	X
Summer vegetables					X
Fall bulbs					
FERTILIZATION					
Trees			X	X	X
Shrubs			X	X	X
Herbaceous plants			X	X	X
PRUNING					
Spring-flowering trees and shrubs			X	X	X
Summer-flowering trees and shrubs			X	X	X
Conifers (candle pruning)					X
Broadleaf evergreens			X	X	X
MULCHING					
	X	X	X	X	X
WEED CONTROL					
Pre-emergence				X	X
Post-emergence					X
TURF					
Establishment				X	X
Overseeding					X
Aerating				X	X
Thatching			X	X	X

do when and past activities you can refer back to. Some gardeners like to use charts, checklists, or even daily garden diaries. Any of these can make your life easier and will help keep you organized.

This is a sample chart showing typical garden activities by season. The types of activities that you would want to record are things such as planting, pruning, fertilizing, mulching, and pesticide applications. This sample chart could be photocopied and used each year. Schedules like this tend to be a bit more general, while a daily diary can be quite detailed.

Late Spring	Early Summer	Midsummer	Late Summer	Early Fall	Mid-Fall	Late Fall
X	X			X	X	X
X				X	X	X
X	X					
				X	X	X
X				X	X	
X				X	X	
X				X	X	
X	X	X				
X	X	X				
X	X					
X	X	X				
X	X	X	X	X	X	X
				X	X	
X	X	X	X	X	X	
X	X			X	X	
X	X		X	X	X	
X			X	X	X	
X						

A SAMPLE GARDEN PLAN

Sustainable gardening gives you so many opportunities to retrofit an existing garden or start over completely with fresh new ideas. Whether you decide to partially or totally renovate your landscape, there are many sustainable initiatives you can implement that will make a difference. The beauty of gardening is that every day is a new one with endless possibilities. If something isn't working for you, it can be changed for the better fairly quickly. We love gardens because we get to witness our ideas growing along with the plants themselves and it inspires us. The sustainable garden allows our ideas and hard work to last for many years to come while improving our quality of life.

Home gardeners can design and implement almost any idea that they can dream up, with few exceptions. Time and money are usually the only two limiting factors, since enthusiasm is usually abundant in most home gardens. The key to choosing the right landscape for you is to make sure your garden meets all your needs and you follow sound gardening practices. A well-designed, balanced garden should offer aesthetics, function, and seasonal interest as well as sustainable qualities such as water conservation, drought and pest resistance, and wildlife habitat.

- Use a balanced combination of tall, medium, and small plants together with a wider variety of blooming times or other ornamental qualities such as fall foliage, fruit, and so on.
- In truly native wooded areas, try to plant only native plants that are indigenous to that area. In a cultivated garden, you can be a lot bolder.

If your goal is to attract and support wildlife, be sure your garden plan includes plants they prefer, along with the resources they need, such as water, nesting habitat, and food.

SETTING REALISTIC GOALS

Gardeners need to know the limitations of their gardens and learn to live within the confines of their resources. I have seen many a gardener who acted like a kid in a candy store, wanting anything and everything to plant in their garden at once. Unfortunately, sound gardening does not work that way, and I have seen real horror shows in which overactive imaginations and well-intentioned enthusiasm went awry. The last thing you want to do is spend a lot of time and money on a garden that is poorly designed, overplanted, and not the least bit sustainable. Besides setting up a garden planner, drawing up a landscape design, and following sound gardening practices, establishing realistic goals is another key piece of the puzzle. As tedious as this might sound, a garden with no plan and no strong set of goals will be extremely difficult to sustain over the long term. The best way to define the goals for your garden is to set up both short-term and long-term sustainable timelines, which will keep you on the right track. Unfortunately, many home gardens today are typically not designed for the long term, but yours must be if you want it to be more sustainable. Timelines can be appropriately set at intervals of 6 months, 1 year, 5 years, 10 years, 20 years, and beyond.

6 Months

The first 6 months of transforming your garden into a more sustainable, low-maintenance garden should really be spent on planning and design. This would also be a good time to get organized and begin the process of choosing materials you want to use in the garden. Evaluating the site, sketching and deciding what should go where, and taking photographs of each area to go along with your plan on paper are prudent steps. Life in the garden starts in the soil, so the first 6 months of developing your new and improved garden should also focus on testing and working the soil to maximize soil health.

Things to Consider for the First 6 Months of Garden Development

- Develop a garden plan on paper.
- Take a soil test and complete soil analysis on nutrients and soil pH.
- Develop a compost program and add to soil as an amendment.
- Based on the soil test, start to add lime, fertilizer, and other materials to boost soil fertility.
- Make sure soil is graded properly and address drainage issues.
- Make a checklist of specific plants or garden features you want to acquire.
- Lay out and mark garden features such as new trees, garden beds, berms, rain gardens, lawn, and so on.
- Start collecting and aging compost and mulch for use over the next 6 months to a year.
- Start to amend and turn over soil in areas designated for vegetables or cut flowers.
- Where you're developing grass meadows or native lawns, kill lawn areas or open spaces by laying down layers of newspaper and compost and plant with a cover crop.

Year 1

While the first 6 months should be used to devise and set up a game plan for the garden, the next 6 months should be a continuation of those efforts. It should also include implementing some of the infrastructure improvements and developing easy-to-establish plantings such as perennial borders, vegetable garden, lawn, and other landscape features that will establish in a short period of time. If your plan was to reduce an already-too-large lawn and expand garden beds, now would be the time to identify those areas and start to bring in soil, compost, and mulch to make it happen.

Things to Consider for the First Year of Garden Development

- Continue composting and incorporating compost into the soil.
- Continue enhancing soil pH and nutrient levels.
- Craft garden beds, raised planters, and berms.
- Reduce the size of existing lawn or renovate or overseed it; add small lawn area using low-maintenance and/or native grass species.
- Begin planting garden beds with herbaceous plants, native ornamental grasses, and shrubs.
- Begin the process of selecting and placing small and large trees.
- Identify and sculpt area to be used as a rain garden or dry streambed to divert water.
- Eradicate invasive weeds in garden beds, lawn, and natural areas.
- Plant native grass in the area you have been clearing or preparing for meadow planting.

Year 5

Now that the easier-to-establish plantings such as herbaceous planting, lawn, native grasses, shrubs, and smaller trees are in place, it is time to think about the long-range planting plan for the landscape. By Year 5, you should really think about establishing vegetative cover such as evergreen screens and the taller canopy of deciduous trees. These are the plantings that will take the most time to establish and grow to the sizes where they can be of the most value. This is also a good time to continue the development and long-term maintenance of flowering shrubs that will hopefully help to create the bones, or structure, of the garden.

Things to Consider for the Next 5 Years of Garden Development

- Continue to plant and establish evergreen and deciduous shrub hedges, screens, and woody plantings to create height and depth in the landscape.
- Plant larger deciduous trees for eventual summer shade.
- Start up a management plan for trees and shrubs, including annual pruning, mulching, and pest management.
- Develop a rotation plan to divide and rejuvenate herbaceous plantings on an annual or biannual basis.

- In natural areas, develop tree and shrub plantings such as holly, viburnum, and dogwood as shelter and a food source for wildlife.
- Continue to mow grass meadows once or twice a year or lawns every few weeks, eradicating invasive species as needed.
- Install and manage birdhouses and bat houses in larger trees or vegetated areas.

Year 10 and Beyond

The key to a successful sustainability program for your garden is the long-range plans and how they are implemented. It is a lot easier to manage the garden over the short term, but having the foresight and vision to consider 10 or more years down the road is both challenging and exhilarating. There are many very positive things that can be done for the first few years to make your garden more sustainable, but it's that ability to continue pushing forward with these successful initiatives for 10 or more years that separates the traditional home garden designed for short-term pleasure from the more sustainable, eco-friendly garden that requires less and offers more benefits in the future. For the first decade and beyond, managing the sustainable garden is all about doing less and reaping more benefits. It may take a while to set up and get going, but if done right, your sustainable garden should require less maintenance and interference as time goes by. By this time the priorities should be to continue to plant and manage larger trees and natural habitat, reduce or eliminate the lawn, replace it with groundcovers as larger trees start to cast shade, continue other low-maintenance initiatives, and refine ideas or initiatives that aren't working as best as they could be.

Things to Consider for the Next 10 Years of Garden Development

- Manage and continue to plant or replace larger trees.
- As tree canopy begins to shade more of the garden, adjust the types and locations of lower plantings to meet those cultural needs.
- Continue regular selective and rejuvenation pruning on trees and shrubs in both cultivated and natural areas.
- Remove or further reduce the lawn area and replace with native plants, groundcovers, and other plants that require less maintenance and more shade.
- Periodically remove or relocate flower beds or shrub borders to meet the needs of your evolving landscape.
- Continue incorporating mulch and compost to maintain organic matter levels in the soil and to promote a healthy soil environment.
- As native habitat continues to flourish, add or manage bat houses and birdhouses.

CHAPTER 2

THE RIGHT PLANT FOR THE RIGHT PLACE

Knowing What Plants to Select and How to Use Them

Properly selecting and siting your bulbs, annuals, perennials, shrubs, or trees is easier said than done. It's quite simple to buy some plants at the local nursery and drop them wherever you have room in the garden. It is much more challenging to determine what you are hoping to accomplish or basing your plant purchases on a specific need and then choosing the right plants for the job. Gardeners buy plants on impulse because the plants look good, and then they bring them home looking for room in the garden for them. It is much more prudent to first evaluate your yard and your specific goals and needs, paying close attention to the environmental conditions that your site offers. Once that is done, then a trip to the local nursery can be more focused on what plants will serve your needs and thrive in the soil, light, and overall climate of *your* garden.

The proper selection and siting of plants in the garden is without question one of the most important and sometimes challenging parts of being a good gardener. Sustainable gardening is more than just planting and caring for a bunch of pretty flowers. It is a passionate quest to grow plants that will reach their greatest potential, maximizing their aesthetic value and function in the landscape. The ultimate satisfaction for gardeners is to watch the plants that they started from tiny seeds develop into beautiful mature plants, knowing that they had a small part in this growth.

SITE ASSESSMENT: THINK FROM THE GROUND UP!

Before you can determine the right plants for your garden, you need to take a close look at what your garden has to offer. There are many environmental factors that go into the proper growth and cultivation of plants. These factors include soil type and pH, light exposure, wind exposure, surrounding vegetation, wildlife considerations, and so forth. Ask yourself these questions about your site:

- What type of soil do you have? Is it clay, loam, or sand? (See Chapter 6.)
- What is the soil pH? Is your soil compacted or in need of amendments? (See Chapter 6.)
- Is your garden in full sun, part sun and part shade, deep shade, or a combination?
- Do you live on an exposed site with frequent wind?
- Does your garden have smaller microclimates or areas that are slightly different in temperature than the rest of the garden?

Concentrate on planting new plants that are suitable for your climate and soil type and planting them properly.

All these questions should be answered before investing a lot of time and money into your garden. By knowing what your garden has to offer, you can determine what plants will grow there. Trying to grow plants that are not suited for your specific climate or environmental conditions will prove to be a frustrating exercise in futility. Many gardeners have tried and failed to retrofit their gardens to suit the needs of plants that are not adapted to grow in that situation. Too much time and money will be wasted trying to get plants to do what we want them to do rather than what they are adapted to do naturally. It is important to realize that some plants can't be grown in certain situations and time can be better spent concentrating on what you can grow rather than what you cannot. That said, sometimes it is good to experiment with new plants. There is, no doubt, a wonderful world of new garden plants just waiting to be discovered. The key is to try out new plants that are rated to grow in your climate and soil type.

Besides assessing the site conditions of your yard, it is also important to seek out well-grown, high-quality plant material. Not that there aren't bargains to be had, but buying plants from a reliable, reputable source is very important. Having a personal relationship with a nursery professional who can get you what you need and also keep you updated on new offerings will prove to be invaluable.

WHAT DOES A HEATHY PLANT LOOK LIKE?

In addition to knowing the source of your plant material, knowing what to look for in the quality of that material is equally important. There are physical attributes that well-grown, high-quality plant material should possess, including healthy foliage, a well-established root system, and well-formed branching crowns and trunks with minimal damage. Following is a checklist of items to take into consideration when shopping for high-quality plant material:

- Plants should show good overall health and vigor, with adequate leaf size (leaves that don't look stunted or misshapen) and good leaf color. Leaves should not be wilted.
- Leaves should not be severely scorched (with burned edges), chewed, or otherwise damaged by insects or disease.
- Branches and trunk on trees and shrubs should not have significant scrapes or wounds.
- Rootballs should be evenly balanced and firm, not lopsided or broken.
- Avoid plants that are top-heavy or severely rootbound.

Nursery plants can come as bare root, in containers, or balled and burlapped.

- Containerized plants should be planted firmly in the container and not wobbly. If you are not convinced the root system of the plant is healthy, ask a nursery salesperson to pull the plant out of the pot gently and make sure the root system is healthy, with whitish root tips.

These simple inspection tips will ensure that you are not buying poor-quality plants. While it's okay to tolerate a few bumps and bruises on plants, significant imperfections will only lead to your valuable plants dying before you have a chance to enjoy them. Poor-quality plants are generally stressed, which means that they take longer to establish and are more susceptible to disease, insect infestations, drought, and root rot.

Selecting annuals and perennials is a bit different, but the same principles apply. These colorful flowers can be sold in cell packs or plugs; may be container, bare root, or field grown; and can be put into pots for transport and sale. In general, container-grown plants should appear healthy and vigorous and not be wilted. Plants should be well rooted and established in the container that they are growing in. This can be determined by gently pulling the plant out of the pot to inspect the root system. A well-established root system should reach the sides of the container and have a firm rootball but not excessive root growth surrounding the inside of the container (potbound). Potbound plants dry out faster, are generally more stressed, and are harder to acclimate to a new garden environment. One of the best ways to help a potbound plant establish is by taking a handheld cultivator and gently teasing the entire rootball so that the roots are no longer growing in a circular pattern. This will stimulate the plant roots to grow into the soil rather than continuously growing in a circle. Last, as a general rule, make sure your plants are not planted too deep. Plants should be planted at ground level or slightly above the grade of the soil.

Potbound plants must have their roots loosened prior to planting to encourage the root system to spread out into the soil.

IT'S ALL ABOUT THE PLANTS

Plants are the backbone of the sustainability movement and are essential to the future of all living things. While this is a pretty bold statement, it is a fact that plants provide food and habitat, prevent erosion, purify air and water, and mitigate climate change. In addition to all of these important benefits, plants provide aesthetic value and increase the property value of your home. The plants recommended in this chapter are adaptable, durable, and functional additions to the landscape that will serve the garden well.

AND THE WINNERS ARE . . .

This section highlights some great garden plants that will enable you to create a more sustainable garden. The selection, siting, and proper planting of superior varieties of plants for the garden is one of the most important aspects of sustainable gardening. Doing all the right things—proper pest management, composting, watering, recycling, and so on—will be for naught if you misuse plants or choose the wrong ones for your specific needs. The key to a successful landscape is to pay careful attention to the proper selection of flowers, shrubs, trees, groundcovers, and vines.

The list below includes species and varieties of some of my favorite plants for almost any landscape situation. These exceptional plants are versatile and reliable performers in the garden and are based on years of observation and admiration in a variety of landscape situations. Both naturally occurring species as well as cultivated varieties have been chosen to provide a complete and diverse offering, and all these plants were selected because of their sustainable attributes. These attributes include environmental adaptability and pest resistance, beneficialness to birds and pollinators, landscape function, and exceptional aesthetic value. Which ones you choose will depend not just on the physical appearance of the plant but also on your climate and geographical region.

SHRUBS

Glossy Abelia (*Abelia × grandiflora*)

Ornamental Value

Masses of small, tubular, pale pink to white flowers cover the plant from early summer until the first hard frost in late fall. It is one of the longest-blooming flowering shrubs available. The dark green, glossy leaves turn rich shades of reddish maroon in fall along with rosy pink flower stalks that persist well after the flowers have fallen off. Abelia will be visited often by desirable pollinators such as bees and butterflies.

Landscape Value

The upright, rounded, dense growth habit can reach 6 feet (1.8 m) tall and equally wide. Glossy abelia is an excellent formal or informal hedge, foundation planting, and companion to perennials, as it plays nicely with its neighbors in a mixed border.

Cultural Requirements

Glossy abelia is heat, drought, and pest tolerant, adapting to a wide variety of soil types and light exposure. It prefers well-drained, moist soil and full sun or partial shade. Pruning is not often needed, but glossy abelia can be sheared into a hedge, selectively pruned, or rejuvenated in early spring. This plant is quick to establish, is pest-free, and requires minimal care once established, making it a very sustainable addition to the garden. It is hardy to temperatures as low as −10°F (23.3°C) to 30°F (−1.1°C)

Cultivars

There are several dwarf forms of glossy abelia that require little or no pruning and not a lot of maintenance. 'Rose Creek' is an outstanding semidwarf variety growing to 3.3 feet (1 m) tall and wide. It offers a profusion of blooms and flowers later in the season than most abelia. 'Little Richard' is another compact form growing up to 3 feet (0.9 m) tall and wide, offering a dense, mounded habit and white flowers. 'Sherwoodii' is a popular variety creeping along the ground and reaching 3 feet (0.9 m) tall and spreading to 5 feet (1.5 m). Many new varieties offer various types of striking variegated and colored foliage.

Related Species

Abelia mosanensis, also known as fragrant abelia, has lustrous, dark green leaves and an upright, arching growth habit to 6 feet (1.8 m). Pinkish white flowers smother the plant in spring and offer an intoxicating sweet fragrance that rivals that of lilac. It is a favorite of butterflies and is also known to be pest resistant. Fall foliage color can range from orange to red. This plant is hardy in temperatures as low as −20°F (−28.9°C).

Bottlebrush Buckeye (*Aesculus parviflora*)

Ornamental Value

Bottlebrush buckeye is a four-season plant displaying dark green, palmlike foliage; 12-inch (30.5 cm)-long, frilly white bottlebrush-like flowers in midsummer; golden yellow fall color; and smooth, gray bark in winter. This plant is interesting regardless of the time of year.

Landscape Value

Bottlebrush buckeye can double as a large shrub or a small tree. It does need room and can reach 8 to 12 feet (2.4 to 3.7 m) tall and 8 to 15 feet (2.4 to 4.6 m) wide. If given ample room, this amazingly durable plant can be used as a specimen, mass planting, or even as a tall screen. Please be advised that this plant is highly poisonous if ingested.

Cultural Requirements

This is a rock-solid plant with very few pest problems. Bottlebrush buckeye prefers moist, well-drained soil and full sun or partial shade. To keep it inbounds in the landscape, occasional selective pruning can be done in early spring to remove older, mature stems. With the exception of this regular pruning, bottlebrush buckeye is relatively easy to maintain, making it quite sustainable. This plant is hardy in temperatures as low as −30°F (−34.4°C).

Cultivars

A naturally occurring variety, *serotina* 'Rogers', blooms a few weeks later than the species and displays larger blooms to 30 inches (76.2 cm) long.

Summersweet Clethra (*Clethra alnifolia*)

Ornamental Value

Summersweet clethra is called that because it blooms in late summer and provides a rich, intoxicating fragrance. Spikes of white (or pink) flowers 3 to 5 inches (7.6 to 12.7 cm) long cover this upright, medium-sized shrub that can reach 8 feet (2.4 m) tall with a slightly smaller width. Leaves are dark green in summer, turning a pale or rich golden yellow in fall. Flowers will attract bees, hummingbirds, and butterflies. Seedpods will hold on for most of the winter, providing a food source for birds.

Landscape Value

Summersweet clethra grows in a wide variety of landscape situations, from moist woodlands to hot, dry seashores. It is an excellent companion plant to other flowering shrubs and herbaceous plants. It should be used in small groupings, informal hedges, or mass plantings.

Cultural Requirements

This tough plant is rather pest resistant and will tolerate poorly drained, heavy soil or sandy, dry soil. It will also tolerate occasional flooding and salt spray. Summersweet clethra will tolerate full sun or dense shade. It's best grown in partial to full sun and moist, well-drained garden soil. This plant's sustainable qualities include versatility, adaptability, and ability to work well with a wide variety of plants. It is hardy in winter temperatures as low as –30°F (–34.4°C).

Cultivars

Many garden varieties of this plant have been introduced over the years. 'Compacta' is a striking variety with a dense habit to 6 feet (1.8 m) tall and wide, very dark green foliage, and white blooms 6 inches (15.2 cm) long. 'Hummingbird' is a popular variety that offers a compact habit to 2 or 3 feet (0.6 to 0.9 m) tall and a spreading habit. 'Sixteen Candles' and Sugartina® are two additional dwarf forms that are excellent choices in gardens with limited space. 'Ruby Spice' displays outstanding deep pink flowers with excellent fragrance and dark green foliage.

Hydrangea (*Hydrangea* spp.)

Ornamental Value

Hydrangeas are beloved around the world and are now one of the most popular and cultivated plants on the planet. With a half a dozen mainstream species and hundreds of cultivated varieties available, hydrangeas are the rock stars of the flowering shrub world. While you wouldn't normally associate hydrangeas with sustainable landscapes due to their need for adequate soil moisture, there are several species—including the Asian native panicle hydrangea (*Hydrangea paniculata*) and North American native oakleaf hydrangea (*Hydrangea quercifolia*)—which I would consider more durable and adaptable than most hydrangea species. *H. paniculata* is a shrub or tree form that flowers later in the growing season compared to most species. Flowers are rounded to cone-shaped white flowers often fading to pink and reach 6 to 8 inches (15.2 to 20.3 cm) long. Fall color can range from yellow to orange and even tinges of red. *H. quercifolia* offers thick, leathery leaves that are shaped like those of an oak tree, which turn rich shades of red and burgundy in the fall. Midseason flowers are also white and typically fade to pink, reaching up to 12 inches (30.5 cm) long. Even in the winter, this plant is interesting with cinnamon-colored, flaking bark.

Landscape Value

Both of these species can be used as single specimens, as screening plants, or in mixed borders.

Cultural Requirements

Panicle hydrangea and oakleaf hydrangea are relatively carefree with few pest problems, adaptability to soil and light, and even tolerance of heat and drought. Panicle hydrangea is hardy to −40°F (−40°C), while oakleaf hydrangea is hardy to −20°F (−28.9°C).

Cultivars

Panicle hydrangea: 'Lime Light' is still my favorite with lime green flowers that mature to pure white. Pinky Winky® and Vanilla Strawberry™ are excellent pink forms. Of the dwarf forms, Little Lime®, 'Little Lamb,' Bobo®, and Babylace® are noteworthy.

Oakleaf hydrangea: 'Alice' with large, pure white flowers and 'Ruby Slippers' with flowers that age to deep ruby red are real showstoppers. 'Sikes Dwarf' is a lovely compact selection with densely branched growth habit. 'Snowflake' is an old-fashioned variety with very elegant and intricate double white flowers turning to rosy pink as they fade. The large, heavy flowers droop off each branch in graceful clusters and change to pink in the fall. Oakleaf hydrangea flowers are wonderful as dried, cut flowers.

Seven-Son Flower (*Heptacodium miconioides*)

Ornamental Value

Seven-son flower can be a large shrub or a small tree growing to 15 to 20 feet (4.6 to 6.1 m) tall and 10 to 15 feet (3 to 4.6 m) wide. This plant is interesting throughout the year, displaying delicate clusters of fragrant white flowers in late summer and lustrous, dark green leaves and bunches of ruby-red flower stalks in fall and early winter. The flowers will attract butterflies and bees to the garden. During the winter the upright, arching branching habit and peeling, tan to silvery-gray bark glisten in the landscape.

Landscape Value

Seven-son flower is effective as a single specimen near the foundation of the house or in a lawn area. It is also an excellent addition to the back of a mixed border as a backdrop to smaller flowering plants.

Cultural Requirements

This adaptable shrub tolerates a wide variety of soil but does best in moist, well-drained soil. It also thrives in full sun and is quite drought tolerant once established. As the plant matures, it is wise to prune off the lower limbs to expose the striking, exfoliating bark, which gets better with age. What makes this plant so sustainable is that once it's established, you can just leave it alone, because it requires minimal care. It is hardy to low temperatures of −20°F (−28.9°C).

Winterberry Holly (*Ilex verticillata*)

Ornamental Value

Winterberry holly is a deciduous holly, meaning it will naturally drop its smooth leaves in the fall and regain them in the spring. Unlike evergreen hollies, winterberry is bare in the winter except for its beautiful, bright red berries in circular clusters along each stem. In the summer, the fairly generic, non-spiny leaves offer a rich, dark green color and turn bright yellow in autumn before falling. Winterberry holly is a real showpiece in the winter landscape. It grows 6 to 10 feet (1.8 to 3 m) in height and spread with an upright and arching branching habit.

Landscape Value

Winterberry holly, like most hollies, is dioecious, meaning female plants need a male plant as a pollinator in order to produce berries. So placing a small grouping of female hollies and one male pollinator someplace close by is a good idea. When you buy your plants, you can ask your nursery professional which are male and female. Winterberry is highly valued by birds, and by my observations, it is eaten early in the winter, while most evergreen hollies are eaten later in the winter. American robins and northern cardinals seem most attracted to this plant. Placing this plant near a window or someplace visible from the house will ensure that you get a glimpse of the wildlife enjoying it as much as you do.

Cultural Requirements

Winterberry naturally grows along streams and rivers in wet soils, but it will also thrive in moist, well-drained, acidic soil and either full sun or partial shade. Pruning can be kept to a minimum, since this plant does not typically grow fast; but occasional selective pruning in late winter will keep plants thriving. Because of winterberry's ability to grow in both wet and dry conditions and even near the seashore, and because it requires very little maintenance, it is a great addition to the sustainable garden. It is hardy to temperatures as low as –40°F (–40°C).

Cultivars

'Red Sprite' is one of my favorites, with noticeable larger, bright red fruit and a semi-compact growth habit. Winter Red™ is a very reliable, heavy fruiting plant. 'Winter Gold' offers large berries that change from green to orange and eventually a golden yellow. 'Autumn Glow' is a hybrid offering a shrubby habit to 10 feet (3 meters) tall, small red fruit that birds love, and yellow fall color. Good male pollinators include 'Jim Dandy' and 'Southern Gentleman.'

Ninebark (*Physocarpus opulifolius*)

Ornamental Value

Ninebark is a North American native that has similar qualities to other flowering shrubs like viburnum and weigela offering ornamental flowers, interesting foliage, and versatility in the landscape. Although mature plants can grow to 5 to 8 feet (1.5 to 2.4 m) high with a graceful, arching habit, many cultivars remain more compact. The white to pink spring showy flowers and leaves that range from green to gold, copper, and purple depending on the cultivar are a beautiful combination. Fall color can range from yellow to bronze in green foliage types. Older plants will exhibit exfoliating bark, which will offer winter interest.

Landscape Value

Ninebark can be used in a variety of landscape situations as an informal hedge, foundation planting, specimen, and mixed border. Often the colored foliage varieties are used in ornamental containers as well.

Cultural Requirements

Ninebark is quite tough, adapting to rocky, dry soils or even clay soils. They are also quite drought tolerant once established. Full sun or partial shade is preferred. This is a rather pest-free shrub and very cold-hardy plant growing in winter temperatures as cold as −50°F (−45.6°C).

Cultivars

This plant has become highly cultivated with a variety of foliage colors and dwarf growth habits that allow it to be more effective in residential gardens with limited space. 'Diablo', Summer Wine®, and Little Devil™ are excellent purple-leaved cultivars, while 'Darts Gold' and 'Nugget' offer gold foliage. Coppertina® starts off with copper-colored foliage that changes to purple, while Amber Jubilee™ offers stunning coppery orange leaves.

Landscape Roses (*Rosa* spp.)

It is important to note that the roses being presented here are not the traditional tea or climbing roses that we know and love so much. Rather, this is a list of landscape roses that offer a shrubby habit, colorful flowers, and most importantly, significant adaptability and disease resistance, unlike tea roses. In short, the landscape roses are quite sustainable in the landscape.

Ornamental Value

Although landscape roses tend not to have quite the same showy, fragrant flowers as tea roses, they still have good color, interesting foliage, and valuable growth habits that can range from low and spreading to upright and hedgelike. Flowers can range from single to semidouble or double and display bright colors, including yellow, orange, red, salmon, and pink.

Landscape Value

Depending on the variety, landscape roses make excellent low, mass plantings and groupings, informal hedges, screens, and companion plants to other flowering shrubs and herbaceous plants. They can even be used in containers in the right situation.

Cultural Requirements

Landscape roses require full sun and well-drained soil. Because they are generally low maintenance, they do not need regular spraying to keep them free of pests and do not require a high-fertilizer diet. Landscape roses are also quite heat and drought tolerant once they are established in the landscape. Pruning is not as intense as that done for hybrid tea roses and usually consists of occasional trimming or shearing to keep plants in check or looking tidy. Avoid significant pruning in summer and fall months. Just the facts that these landscape roses require no pesticides to keep them healthy and need less intense pruning than the more traditional roses make these plants a must-have in any eco-friendly garden. Landscape roses are hardy to –20°F (–28.9°C).

Cultivars

Knock Out® roses were developed specifically for landscape function and disease resistance. They come in a wide variety of colors, including Blushing Knock Out®, Double Knock Out®, Pink Knock Out®, Sunny Knockout®, and the original Knock Out®. Established plants can range in size from 3 to 4 feet (0.9 to 1.2 m) tall and wide. If plants get too big or are not performing well, they can be pruned down to 12 inches (30.5 cm) while dormant in later winter or early spring and will bounce right back the same growing season with lush foliage and gorgeous flowers. I have witnessed these roses looking great in many places all over the world, from hot and humid to cold and dry.

But there are other good shrub roses available that can function as groundcovers, foundation plantings, and mass plantings and as companions to herbaceous plants. Flower Carpet® roses are another outstanding series that offer low, ground-hugging roses growing under 3 feet (0.9 m) in height and 3 to 4 feet (0.9 to 1.2 m) in width. The Flower Carpet® roses have small, showy blooms in clusters and glossy, dark green foliage. They are especially known for their pest resistance and heat and drought tolerance. These plants were even tested in Australia and passed the test

for durability and tolerance to extreme heat! They will grow in warmer climates but also in climates that reach lows of −20°F (−28.9°C). In warmer climates, it is advisable to grow these roses in part shade. Flower colors include white, light pink, coral, deep rosy pink, yellow, and red. Plants can be cut down in late winter or early spring and rejuvenated, flowering in the same season.

A few other good roses to complement the sustainable garden include the low-growing Drift® series and the shrubby Carefree® series. Both groups offer a wide variety of flower types and colors as well as adaptability in the landscape. Most important, they do not need the regular pesticide applications, fertilizer, or detailed pruning that conventional roses require to keep them looking good.

Littleleaf Lilac (*Syringa microphylla* 'Superba')

Like roses, one of the most beloved flowering shrubs, lilacs have experienced a renaissance. The common lilac is amazingly fragrant, but once it's finished blooming, it is a rather unattractive plant. By late summer, it often gets a white haze on its leaves known as powdery mildew. Several other species and varieties of lilac are more landscape friendly, offering a smaller, compact habit and much better resistance to diseases. Although not as large and showy, the flowers are still colorful and fragrant. Littleleaf lilac and others are worth including in the landscape as informal hedges, groupings, and foundation plantings.

Ornamental Value

Littleleaf lilac displays rosy pink flowers in spring and often reblooms in late summer or autumn. The shrub has small, delicate leaves and grows to 6 feet (1.8 m) tall and 12 feet (3.7 m) wide, although it is a slow grower.

Landscape Value

Littleleaf lilac develops into a dense shrub with cascading branches. It is effective as an informal hedge, foundation planting, or backdrop to herbaceous plants and other flowering shrubs. Because of its graceful habit and profuse display of blooms, littleleaf lilac is more user friendly in a smaller, residential garden than the common lilac is. Your garden will be buzzing with activity during the spring with all the butterflies, bees, and hummingbirds that will visit your lilacs.

Cultural Requirements

Like most lilacs, littleleaf lilac prefers moist, well-drained soil and full sun or partial shade. It is very important that the soil drains well; otherwise, root rot can be a problem. Pruning needs are minimal, and occasional selective pruning in late winter to remove older, less productive stems will keep plants more compact and floriferous. This species is quite resistant to common problems that typically plague lilacs, such as powdery mildew. These dwarf lilacs need a few years to establish, and then you can just sit back and watch them flourish. This plant is also tolerant of heat and humidity and is also quite cold tolerant, surviving in winter temperatures to −30°F (−34.4°C).

Related Species

Meyer lilac (*Syringa meyeri* 'Palibin') is another landscape-friendly lilac with violet-purple flowers; small, glossy leaves; and a compact habit to 4 feet (1.2 m) tall and 6 to 7 feet (1.8 to 2.1 m) wide. This plant makes a very nice hedge or grouping in areas where you don't have a lot of room. It is hardy to temperatures as low as −40°F (−40°C).

The Fairytale® series is a wonderful group of dwarf lilacs that offer various shades of pink flowers, a dense habit, and lustrous foliage. Fairy Dust®, Prince Charming®, Sugar Plum Fairy®, Thumbelina®, and Tinkerbelle® are all unique and beautiful in their own right. They are disease resistant and are especially good for colder climates, as they will survive in climates with winter temperatures as low as −40°F (−40°C).

VIBURNUM (*Viburnum* spp.)

Asking horticulturists which viburnum they like best is like asking parents which kid is their favorite. Viburnums are so diverse and versatile in the garden, it's impossible to pick just one to use. In fact, you could landscape your garden just using different species and varieties of viburnum and never be bored. That said, here are a few of my favorites that offer outstanding ornamental value, unrivaled landscape function, and food for wildlife. The flowers of these viburnums will attract bees and butterflies, and the fruit will provide a nutritious feast for birds in winter. For the most reliable fruit display, plant viburnums in groups so there is good cross-pollination. Viburnums are like potato chips—you can't just have one! For the gardener who wants to have a more sustainable and beautiful garden, viburnums are a dream come true. Because they provide color all season, provide food for wildlife, and are adaptable to many landscape situations, viburnums are a gardening treasure.

Korean Spicebush Viburnum (*Viburnum carlesii*)

Ornamental Value

Korean spicebush viburnum offers rounded flower clusters that are pink in bud and pure white when open. The flowers are intoxicatingly fragrant and rival the potency of lilac. The entire landscape will smell of the beautiful perfume that this plant offers for several weeks. The fuzzy, dark green leaves turn rich shades of red in fall, and the growth habit is dense and rounded to 4 to 6 feet (1.2 to 1.8 m) tall and wide. In late summer and fall, red fruits turn to black and are not conspicuous but certainly edible to birds.

Landscape Value

Korean spicebush viburnum is an excellent informal hedge or small grouping. It is a fine choice to place near the foundation of the house or a walkway so you can enjoy the fragrance of the flowers. The large, white flowers are beacons to butterflies and other beneficial pollinators.

Cultural Requirements

This flowering shrub prefers moist, well-drained soil and full sun or partial shade but is adaptable. It is fairly carefree and does not require pruning very often. It is hardy as cold as −30°F (−34.4°C) with special care.

Cultivars

'Compactum' is a dwarf variety growing only to 3 to 4 feet (0.9 to 1.2 m) tall and wide.

Linden Viburnum (*Viburnum dilatatum*)

Ornamental Value

This is another showy viburnum with outstanding floral, foliage, and fruit display. The linden viburnum shows off large, creamy white flowers that look like miniature Queen Anne's lace flowers. They are perfect landing spots for bees, butterflies, and other helpful insects looking for nectar. The rounded, dark green, textured foliage turns brilliant shades of red and maroon in fall alongside clusters of bright red fruit that look like miniature cranberries. At maturity, plants become dense and rounded, growing 8 to 10 feet (2.4 to 3 m) tall and wide. With judicious pruning in late winter or early spring, plants can be kept smaller.

Landscape Value

Linden viburnum will work well in a shaded woodland garden or in a mixed border with herbaceous plants. It's a great informal hedge, foundation planting, or single specimen.

Cultural Requirements

Linden viburnum is easy to grow, thriving in full sun or partial shade and moist, well-drained soil. It is hardy to temperatures reaching −30°F (−34.4°C) and in warmer climates will benefit with some cool shade in the afternoon.

Cultivars

Cardinal Candy™ is a beautiful form with lush, glossy leaves and large clusters of glossy red fruit. 'Erie' is a vigorous selection with large flowers 4 to 6 inches (10.2 to 15.2 cm) wide and showy red fruit. 'Michael Dodge' offers unusual bright golden yellow fruits that are real eye-catchers in the landscape.

Arrowwood Viburnum (*Viburnum dentatum*)

Ornamental Value

This Native American viburnum is a tall shrub growing 6 to 8 feet (1.8 to 2.4 m) tall and wide, but it can grow larger in ideal situations. The growth habit is upright, dense, and arching, giving this plant a strong and graceful presence in the landscape. The large, flat-topped, white flowers in spring are not at all fragrant but offer a nice display. The sharply serrated, dark green leaves will turn shades of yellow, red, or purple in fall. Bluish black fruit also ripen in fall and offer a tasty treat for birds.

Landscape Value

Arrowwood is most effective in a shaded woodland garden or naturalistic setting. It can be grown in small groupings or as a tall screen or informal hedge.

Cultural Requirements

Arrowwood thrives in partial shade with plenty of drainage and moisture. It will grow in full sun as long as you provide mulch and water it regularly during drought. This species is particularly susceptible to an insect called viburnum leaf beetle (VLB), which will chew up the leaves. Natural predators such as lady beetles (ladybugs) and lacewings will control VLB. Arrowwood viburnum is hardy to −40°F (−40°C).

Cultivars

Blue Muffin™ is a semidwarf variety growing 5 to 7 feet (1.5 to 2.1 m) tall and displaying bunches of bright blue fruit in fall.

Smooth Witherod Viburnum (*Viburnum nudum*)

Ornamental Value

This underutilized viburnum has showy, creamy white, flat-topped flowers in spring, which attract various species of butterflies. The smooth, lustrous, dark green leaves turn brilliant shades of maroon to reddish purple in fall. The clusters of fall fruit turn from pink to blue to purplish black and are a beautiful complement to the red foliage. This upright-growing viburnum can reach 6 to 12 feet (1.8 to 3.7 m) in height with a similar spread.

Landscape Value

Witherod viburnum is an excellent choice for a small grouping, woodland garden, or informal hedge or for a low-lying area that collects rainwater, such as a rain garden.

Cultural Requirements

Witherod viburnum can tolerate a wide variety of soil but prefers moist, loamy soil and will also grow in boggy conditions. It will thrive in well-drained, rich garden soil as well. It prefers full sun or partial shade. Selective pruning to remove older stems in late winter or early spring every few years will keep plants dense and vigorous. It is hardy to −20°F (−28.9°C).

Cultivars

'Winterthur', named after the great garden in Delaware, offers abundant blue fruit and red foliage in the fall. The growth habit is semi-compact to 6 feet (1.8 m) tall and wide. Brandywine™ is a newer selection known as a heavy fruit bearer with a compact habit to 5 feet (1.5 m).

Cranberry Bush Viburnum (*Viburnum trilobum* and *V. opulus*)

Ornamental Value

Cranberry bush viburnum is a large, upright shrub with graceful, arching branches growing 8 to 12 feet (2.4 to 3.7 m) tall with a similar width. Delicate, white, flat-topped flowers emerge in spring followed by large, translucent, bright red berries in fall and winter. The lush, maple-like leaves transform from dark green to yellow, bright red, or deep maroon. The fruits are a bit stinky as they age, so siting this plant away from a patio or walkway is wise.

Landscape Value

Cranberry bush viburnum works best in small groupings or as an informal hedge or screen. Its sturdy branches are the perfect place to hang a bird feeder.

Cultural Requirements

As with most viburnums, moist, well-drained soil and full sun or partial shade are best for flower and fruit production. Occasional selective pruning to remove old stems in early spring is recommended. It is very cold hardy to –50°F (–45.6°C) and will tolerate warmer climates with some shade and extra care.

Cultivars

'Compactum' is a good dwarf form with beautiful flowers and fruit. It grows only to 6 feet (1.8 m) tall and wide, making it ideal for smaller home gardens.

Red Buckeye (*Aesculus pavia*)

Ornamental Value

This southeastern American native tree provides quite a floral display in spring with long spikes (6 to 10 inches [15.2 to 25.4 cm]) of deep red flowers. The individual tubular flowers are great at attracting hummingbirds. The dark green, glossy, palm-shaped leaves and dense, upright habit to 15 to 20 feet (4.6 to 6.1 m) tall are also very attractive. In the late summer and fall, fruit capsules open to expose a chestnut-like fruit, but don't be fooled. This fruit is not edible, even for some wildlife.

Landscape Value

Red buckeye is a very useful single-specimen tree in a lawn or in small groupings. It can also work well in a woodland area with naturalistic plantings. Red buckeye can be grown as a single-stemmed tree or multistemmed large shrub.

Cultural Requirements

Red buckeye prefers full sun or partial shade and moist, well-drained soil. It is quite heat and humidity tolerant and in hot climates will benefit from afternoon shade. It is hardy to temperatures as low as –30°F (–34.4°C).

Shadblow Serviceberry (*Amelanchier canadensis*)

Ornamental Value

Shadblow serviceberry is a tall, upright native tree or large shrub with delicate, small bouquets of white flowers in spring, medium to dark green leaves in summer, and brilliant fall color. Leaves turn various shades of yellow, orange, and red. The smooth, sinuous stems are silvery gray, making them quite noticeable in winter. But probably the most alluring attribute of this plant is the blueberry-like fruits that develop in spring and ripen by early summer. Small fruits change from green to red and eventually blue when they are ripe. If you can beat the birds to them, you'll find that the ripe fruits are sweet and juicy and rival the flavor of blueberries. If not, the birds will feast on them early and often until they are gone. Shadblow serviceberry is usually multistemmed and shrubby, growing between 6 feet (1.8 m) and 20 feet (6.1 m) tall, depending on where they are planted and how they are pruned.

Landscape Value

Shadblow serviceberry is very effective in groupings in a shaded woodland garden where there is dappled light and plenty of air circulation or as a single specimen in a mixed border. It fits in best in a naturalistic setting where it can grow freely.

Cultural Requirements

Because shadblow serviceberry is in the rose family, rust, leaf spots, and other diseases common to roses can occur. But if serviceberry is sited in an open, airy location, diseases can be minimized. This is a durable plant that thrives in moist, well-drained soil but also tolerates hot, dry locations in sandy soil. Full sun or partial shade is preferred, and pruning is not required very often. Selective pruning to remove older or damaged stems can be done in later winter or early spring. Serviceberry is hardy to −40°F (−40°C) but will grow in warmer climates as well with protection from the hot afternoon sun.

Cultivars

A related species, *Amelanchier × grandiflora*, offers good selections with brilliant fall color and tasty fruit, including 'Autumn Brilliance,' 'Autumn Sunset,' and 'Ballerina.'

Eastern Redbud (*Cercis canadensis*)

Ornamental Value

Eastern redbud is an important North American native tree offering bright pink, pea-like flowers in early to mid-spring and lustrous, dark green, heart-shaped leaves all summer. The flowers are not quite red in bud but more of a reddish purple opening to a rosy pink that illuminates the landscape. Even from a distance, this plant is quite noticeable in flower. Flower buds will form in profusion all along the stems and even the main trunk of the tree. The dark green leaves offer a bold texture in the summer but have no significant fall color. Redbud can reach 20 to 35 feet (6.1 to 10.7 m) tall with a similar width at maturity.

Landscape Value

Eastern redbud is an excellent single specimen in the lawn or in small groupings in a woodland setting. You will often see it growing along highways and roadsides. Butterflies and bees like the small, pea-like flowers of redbud.

Cultural Requirements

Eastern redbud fixes atmospheric nitrogen, meaning it can process nitrogen from the air and make its own fertilizer. That means it will grow in barren or low-fertility soils. Although there are quite a few insects and diseases that bother redbud, it is still worth planting—especially some of the new and exciting garden varieties. The key to success with redbud is giving it very well-drained soil and either full sun or partial shade. Too much shade and your plants won't be happy. But overall, because eastern redbud will grow in tough conditions and still flower reliably, it is an excellent addition to a sustainable landscape. It is hardy to −30°F (−34.4°C) and possibly colder with some protection.

Cultivars

This plant has so many great varieties, it is hard to pick just a few. 'Appalachian Red' is the closest to a true red, with neon reddish pink flowers; it will stop you in your tracks. 'Forest Pansy' has rich, luxurious, deep purple leaves that are very striking in the summer landscape. After a fresh rain, beads of raindrops glisten from the colorful foliage. Lavender Twist™ is a showstopper with a strongly weeping habit, masses of pink flowers, and large, lush green leaves that look like a waterfall of foliage in summer. 'Royal White' is one of several "white" redbuds that are quite attractive, especially in partial shade. The Rising Sun™ is one of several new yellow-leaf *Cercis* species with new foliage emerging golden orange before fading to chartreuse and finally green. The talk of the town these days is one of the newest selections called Flame Thrower®, which offers striking reddish burgundy new growth that fades to shades of yellow and then green, often displaying multiple foliage colors at the same time.

Flowering Dogwood (*Cornus florida*)

Ornamental Value

There is nothing quite like a flowering dogwood in full bloom. It doesn't seem like spring officially begins until dogwoods are in full display. The floral display of dogwood is rather complex and is not like many other flowers. The colorful part of each inflorescence, or flower cluster, is called a bract, which is actually a modified leaf. Dogwood blossoms have four white or pink bracts that surround an inconspicuous cluster of flowers. Once pollinated, the flowers eventually transform into beautiful, glossy red clusters of fruit in fall. Dogwood fruits are highly sought after by birds, which will often strip trees clean of their fruit before the onset of winter. In addition, flowering dogwood has brilliant, colorful fall foliage, with leaves typically changing to variations of orange, red, and maroon. The fall color of dogwood rarely disappoints and is consistently showy year after year. Flowering dogwood also develops rough, gray bark like alligator skin. It has a rounded, spreading growth habit reaching 20 to 30 feet (6.1 to 10.7 m) tall and wide.

Landscape Value

Flowering dogwood is an excellent specimen tree in a lawn area or in small groupings in a woodland setting. Having an assortment of colors sprinkled across the landscape will liven up any garden. Flowering dogwood is truly a classic Native American tree and one of my favorites.

Cultural Requirements

Flowering dogwood is known to be susceptible to several damaging pests, including anthracnose and powdery mildew. But susceptibility can be greatly reduced by siting your dogwoods on the east side of the garden in full sun or partial shade. This will allow plants to enjoy morning sunlight and some protection from the hot summer sun in the afternoon. Moist, organic, well-drained, acidic soil is preferred. Mulching trees with a light layer of wood chips and avoiding overhead watering in the afternoon and evening hours will also reduce stress and alleviate disease problems of this plant. It is hardy to –20°F (–28.9°C) and rather adaptable to warmer climates as well.

Cultivars

'Appalachian Spring' is a white variety with good disease resistance and tolerance of heat and humidity. 'Cherokee Princess' offers large, pure white flowers, good vigor, and good disease resistance. 'Cherokee Brave' is a striking variety with deep pink—almost red—flower bracts and excellent red fall color.

Over the past few decades, researchers at Rutgers University in New Jersey have developed hybrid dogwoods that are more disease-resistant than *Cornus florida* but have similar landscape attributes. These hybrids offer large, showy flowers; clean foliage; and slightly later blooming. Aurora®, Celestial™, Ruth Ellen®, Stellar Pink®, and Venus™ are a few selections from the Stellar® series that are excellent additions to the garden.

Crape Myrtle (*Lagerstroemia indica*)

Ornamental Value

Crape myrtle is a popular shrub or small- to medium-sized tree native to the Asian continent. It displays white, pink, red, or purple flowers, depending on the variety selected, in summer and fall. The large panicles (branched clusters of flowers) of crinkled, crepe paper–like flowers develop into clusters of round seedpods that persist into winter. The foliage is dark green in the summer, changing to brilliant shades of yellow, orange, and red in the fall. But probably the most attractive feature of crape myrtle is its smooth, flaking bark that can range in color from gray to tan or reddish brown. This feature is striking all year but is especially noticeable in winter. Crape myrtle can range in size from a 4- to 6-foot (1.2 to 1.8 m) shrub to a 15- to 25-foot-(4.6 to 7.6 m) tall tree.

Landscape Value

Crape myrtle will be visited by bees and other flying insects but is more known for its aesthetic qualities and durability in the landscape. Shrubby varieties make excellent foundation plantings, small groupings, or companions to other flowering shrubs or herbaceous plants.

Cultural Requirements

Without question, crape myrtle is one of the most heat- and drought-tolerant plants available to gardeners. It is rather low maintenance once established and epitomizes the word *sustainability*. Crape myrtles flower on new growth, so pruning is often done in the spring while plants are still dormant. In warmer climates, pruning right after blooms fade may encourage a second flush of flowers. Pruning large shrubs back to very thick stems or main trunks should be avoided, since this may encourage fleshy, wispy growth to develop. Instead, you can thin out the canopy of the plant, and the tips of the branches can be cut back to branches no thicker than your pinky finger. To train taller-growing crape myrtle varieties as small trees, prune off any young, spindly, or thin branches from the lower part of the plant, leaving several mature main trunks. This will also expose the beautiful exfoliating bark.

Crape myrtles adapt to most soils but thrive in well-drained, moist soil and full sun. Partial shade is acceptable, but too much shade should be avoided. They are hardy to –5°F (–20.6°C) but are much happier in warmer climates.

Cultivars

There are countless garden varieties of crape myrtle available on the market. 'Pocomoke' is a dwarf variety offering deep pink flowers and a dense growth habit to 3½ feet (1.1 m) tall and wide. Similar compact varieties such as the Razzle Dazzle® series only grow 3 to 5 feet (0.9 to 1.5 m) wide and tall and require very little maintenance once established. Cherry Dazzle®, Dazzle® Me Pink, and Snow Dazzle® are several good selections. 'Natchez' is a large-growing crape myrtle with white flowers and cinnamon-brown bark that works great as a small tree. The First Editions® magic series offers intermediate-sized crape myrtles that come in a wide variety of flower colors, several of which also feature deep purple foliage that makes a striking contrast to the beautiful flowers.

Crabapple (*Malus* spp.)

Ornamental Value

Crabapples, which are close relatives to more commonly eaten apples, are flowering trees with miniature fruit that offer ornamental value to the landscape as well as feed wildlife (and people; they are often used in jellies). The white, pink, or red fragrant flowers of crabapples provide sweet nectar to bees and other pollinators and provide gardeners with a colorful show for several weeks in the spring. The dense, oval to rounded growth habit and dark green leaves also offer aesthetic value in the landscape. The small apples that range in color from yellow and bright red often persist through the late summer and into the fall and winter months. It is very important to note that in some environments, some species and varieties of crabapple can be invasive. Gardeners should educate themselves before planting a crabapple and avoid crabapples that will pose a threat to the cultivated and natural landscape.

Landscape Value

Crabapples are excellent, durable, adaptable trees that can be grown as single specimens or in groupings. They also work in a lawn area, but I recommend avoiding the placement of crabapples near a walkway, driveway, or patio because of fruit drop.

Cultural Requirements

Crabapples have come a long way from the old-fashioned types that you may be familiar with. Diseases that would distort fruit and leaves and cause plants to defoliate prematurely plagued the old-fashioned types more often than not. But now there are many new cultivated varieties of crabapples that have emerged as tough, disease resistant, and quite user friendly in the garden. In general, crabapples prefer moist, well-drained soil and full sun or partial shade; but frankly, they will adapt to a wide variety of environmental situations. Pruning is tricky with crabapples because they develop vegetative suckers from the roots and water sprouts from the stems that have to be removed regularly in order to keep plants floriferous and looking good. The best time to prune out this pesky, vegetative growth from the ground or cluttering up the canopy of the tree is in midsummer after flowering. Pruning in spring will only stimulate the plant to grow more suckers and water sprouts the next year. Pruning one-third of your crabapple each year over a 3-year period will reduce shock to the plant and also discourage new growth. Crabapples are hardy and also rather heat and drought tolerant in warmer climates. Generally they can survive in climates with winter temperatures down to −30°F (−34.4°C).

Cultivars

Gardeners should do some research and select crabapples that are noninvasive or sterile. I recommend 'Callaway,' which has white flowers and large reddish maroon fruits that are actually tasty to humans. 'Red Jewel' has beautiful white flowers and bright cherry-red fruit. Sugar Tyme® offers masses of pure white flowers and candy apple–red fruits that persist all winter. All these selections offer good resistance to disease and perform well in the landscape.

Ornamental Cherry (*Prunus* spp.)

Ornamental Value

Cherries provide the garden with colorful spring flowers, interesting foliage, and wonderfully textured bark. However, many ornamental cherries just get too large and cumbersome for the average residential landscape. A home garden can be quickly consumed by a fast-growing, wide-spreading cherry. But there are several semidwarf varieties that are manageable with petite, single to semidouble, rosy pink, light pink to nearly white flowers in profusion in spring. The dark green, delicate leaves turn shades of orange and red in the fall. The bark is typically light gray to reddish brown and most noticeable in winter. Plants can be trained as a single stem and are often grafted onto other species of cherry, or they can be multistemmed. Like crabapples, cherries provide sweet nectar for pollinators.

Landscape Value

The selections of flowering cherries listed below are smaller in stature than most cherries and are ideal as specimens for a home garden and can also be used effectively to line a walkway or driveway, in groupings, or to complement a mixed border.

Cultural Requirements

Moist, well-drained soil and full sun or partial shade are best. Although cherries are susceptible to many pest problems, this cherry is rather durable and relatively easy to grow. The flowering cherries listed are hardy to –10°F (–23.3°C) but will survive to colder climates with some protection.

Cultivars and Related Species

'Okame' is another garden hybrid that is also appropriate for the home garden. 'Okame' offers rosy pink flowers in early spring and an upright, vase-shaped growth habit to 20 feet (6.1 m) tall. 'Hally Jolivette' will grow 12 to 15 feet (3.7 to 4.6 m) in height with about half the spread, making it ideal for a home garden *Prunus mume*, also known as flowering apricot, will bloom as early as midwinter or early spring with bright pink flowers. This tough tree thrives in warm climates. 'Peggy Clarke' is a popular variety with double, deep rose flowers; and 'Matsurabara Red' has striking double, dark red flowers.

Japanese Stewartia (*Stewartia pseudocamellia*)

Ornamental Value

Japanese Stewartia, an Asian native, has four seasons of interest. Year-round, this low-maintenance tree will provide continuous ornamental value as few other trees will. The dark green, lush leaves provide a nice backdrop to the large, round flower buds, which look like pearls before they open to pure white circular flowers with yellow centers. Bees will frequent the tree and pollinate the flowers, which will become small brown capsules. This midsummer bloomer will continue to flower for about a month in the landscape. The leaves turn brilliant shades of orange, red, or maroon in the fall. But probably the most identifiable characteristic of this tree is the smooth, multicolored, exfoliating bark, which offers variations of brown, beige, and gray. This beautiful bark feature, along with the tree's strong, upright habit reaching 20 to 35 feet (6.1 to 10.7 m) tall with a slightly smaller width, makes this tree a real standout in the winter landscape.

Landscape Value

Japanese stewartia is the quintessential landscape specimen tree. It is ideal in a lawn, mixed border, or woodland garden. This shining star will get better with age and make everything around it look better.

Cultural Requirements

Japanese stewartia is quite adaptable and carefree, but moist, well-drained soil is recommended. Hot, dry, and exposed sites should be avoided. It is a pest-free plant that, once established, requires little pruning, fertilizer, or special care. Japanese stewartia prefers full sun or partial shade. It takes a few years to establish once planted, so be patient. Like a fine wine, your stewartia will only get better with age. It is hardy to temperatures as low as −20°F (−28.9°C) and possibly a bit colder with some winter protection.

Related Species

Stewartia ovata (mountain Stewartia) and *Stewartia monadelpha* (tall stewartia) are two species that can offer slightly different qualities in the landscape. Mountain stewartia displays larger white flowers with yellow to purple centers. Mountain stewartia is native to the eastern United States. Tall stewartia, an Asian native, has smaller leaves and white flowers; but the upright habit and cinnamon-brown bark are quite striking. Mountain stewartia is one of the most heat-tolerant species of stewartia available. These species can tolerate a bit more heat and humidity and will grow in warmer climates.

Japanese Tree Lilac (*Syringa reticulata*)

Ornamental Value

Although related to the shrubby, fragrant lilacs, this small- to medium-sized tree looks quite different and blooms about a month later than common spring-flowering lilacs. The puffy, large, creamy white flower clusters are not quite as fragrant as other lilacs; but the floral display and other ornamental characteristics surely make up for that. Japanese tree lilacs also display lustrous, dark green foliage in summer and striking, dark, reddish brown, cherry-like bark. Even in youth, this upright grower displays a vase shape with a single trunk that you will not see in most lilacs. Mature specimens can get 20 to 30 feet (7.6 to 9.1 m) tall with about half the spread.

Landscape Value

Japanese tree lilac is excellent as a single specimen or in small groupings. As long as you give this plant the room it needs, it will be a good performer and a carefree addition to the landscape.

Cultural Requirements

Japanese tree lilac is fairly easy to grow, doing best in full sun or partial shade and moist, well-drained soil. However, it is quite tolerant of heat, humidity, drought, and poor soil. It is also resistant to many of the pest problems that plague many other lilac species. This is a cold-hardy species tolerating winter temperatures as low as −40°F (−40°C).

Cultivars

Typically tree lilacs need time to establish before they will flower reliably; however, 'Ivory Silk' is a heavy-flowering form that blooms at a young age with silky white flowers.

HERBACEOUS PLANTS

Common Yarrow (*Achillea millefolium*)

Ornamental Value

Common yarrow is a popular perennial that is grown in many gardens across the world. It has flat-topped flowers that range in color from white to cerise. The fernlike, delicate foliage gives this plant an open, airy look; and flower stalks reach 2 to 3 feet (0.6 to 0.9 m) tall in mid- to late summer. The flowers are perfect landing pads for butterflies and bees.

Landscape Value

Common yarrow is rather aggressive, and although it is not native to the United States, it has naturalized along roadsides and in other native habitats. It is best used in areas where it can't spread too far, since it has a thick, matted growth habit. It is usually found in mixed borders planted along with other summer-blooming perennials.

Cultural Requirements

Common yarrow is best grown in well-drained, sandy loam and full sun. It will tolerate poor soil as long as there is good drainage. Yarrow tolerates heat and humidity but will tend to be top-heavy and flop in areas where warm, humid nights are common. Cutting back plants in early spring will help keep them compact. They can also be cut back after initial flowering, which will often encourage reblooming. Common yarrow, once established, should be divided every few years to keep the plants vigorous and to check their spread. They will form sustainable colonies in areas where left unchecked. Yarrow is generally hardy to temperatures as low as −30°F (−34.4°C) and will thrive in warmer climates as well.

Cultivars

The Seduction™ series have shown to be good performers, including 'Peachy Seduction' and 'Sunny Seduction.' 'Pomegranate' offers rich red flowers and gray-green foliage. Two good yellow garden hybrids worth adding to the garden are 'Coronation Gold' and 'Moonshine.'

Japanese Anemone (*Anemone × hybrida*)

Ornamental Value

Japanese anemone is an excellent late-season bloomer with white or pink flowers 2 to 3 inches (5.1 to 7.6 cm) in diameter emerging in late summer and fall. The rich green, dissected leaves and mounded habit are also attractive. The only drawback of this plant is that many selections grow 3 to 4 feet (0.9 to 1.2 m) tall and need staking, or they will flop all over the garden. But the pros far outweigh the cons with this plant, and it will add much-needed color late in the season. It will also function as a bee magnet, as pollinators love it.

Landscape Value

Japanese anemone is an excellent addition to a mixed-flower border or in groupings. I have seen this plant combined with other earlier season perennials, adding contrast and interest to the garden. Anemones are known to be deer resistant, which is an added bonus of using these outstanding plants.

Cultural Requirements

Japanese anemone prefers moist, well-drained soil and full sun or partial shade. Avoid wet soil and keep this plant well-watered during times of drought. Divide large plants in spring every few years to control their spread. It is hardy to –30°F (–34.4°C).

Cultivars

'Honorine Jobert' is an old-time variety that has been around since the Civil War and is still popular today. It features beautiful white flowers and a robust growth habit 3 to 4 feet (0.9 to 1.2 m) tall. 'Margarete' displays deep pink, semidouble to double flowers and will grow 2 to 3 feet (0.6 to 0.9 m) tall. 'Queen Charlotte' displays large, beautiful, pink semidouble flowers and grows to about 3 feet (0.9 m) tall. 'Whirlwind' is a tall grower to 4 to 5 feet (1.2 to 1.5 m) with large, white semidouble flowers.

Butterfly Weed (*Asclepias tuberosa*)

Ornamental Value

Butterfly weed, as the name suggests, is a real favorite of many species of butterflies and has vibrant, bright orange flowers in spring and summer. The flowers can also have variations of yellow or even red in the flower. Flowers will give way to ornamental seedpods and silky seeds and are often dried and used in floral arrangements. Mature plants can range in size from 1 to 2½ feet (0.3 to 0.8 m). Adult butterflies value the nectar, and caterpillars will feed on the foliage, giving this useful plant a dual purpose.

Landscape Value

Butterfly weed is found growing extensively in many parts of its native North America. It is an excellent choice for a sunny mixed border, meadow, or prairie with other wildflowers or in groupings with other insect-attracting native plants.

Cultural Requirements

Butterfly weed prefers drier soil that drains well, and it tolerates drought quite well due to its tuberous, thick root system. It also thrives in full sun, and it can often be found growing in open fields. It is hardy to −40°F (−40°C) but will also thrive in warmer climates with moderate winter temperatures.

Threadleaf Coreopsis (*Coreopsis verticillata*)

Ornamental Value

Threadleaf coreopsis is native to the eastern United States and is one of the most versatile and adaptable perennials for the garden. The lacy, fine foliage and showy yellow flowers in summer make this plant a standout in the landscape. Established plants will grow 18 to 24 inches (45.7 to 61 cm) tall and develop into large clumps in the garden. Threadleaf coreopsis is excellent for attracting pollinators into the garden.

Landscape Value

Threadleaf coreopsis is an excellent companion plant to other summer-blooming perennials and annuals. It works well in groupings in a mixed border as well. Threadleaf coreopsis is known to be a good, reliable deer-resistant plant in the landscape.

Cultural Requirements

Threadleaf coreopsis prefers full sun, but light shade is acceptable as well. Plants will establish rather quickly in the landscape, forming a sizeable clump growing to 2 to 3 feet (0.6 to 0.9 m) wide by the end of the summer. These clumps can be divided in fall or early spring. Once the initial flush of flowers subsides, plants can be deadheaded and will often rebloom in the fall. This plant is hardy to −40°F (−40°C) and also thrives in moderate climates.

Cultivars

'Moonbeam' is a very popular garden variety with soft yellow flowers. It grows up to 2 feet (0.6 m) tall but will flop when it reaches maturity. 'Zagreb' is a more compact variety that tends to stay upright with bright, golden yellow flowers. 'Full Moon' is a hybrid with large bright yellow flowers 2 to 3 inches (5.1 to 7.6 cm) in diameter. It is a real showpiece in the garden. It is one of several in the Big Bang™ series.

Purple Coneflower (*Echinacea purpurea*)

Ornamental Value

Purple coneflower is a very popular perennial with brown to bronze, cone-shaped flowers surrounded by rose-purple petals in summer. The bold, hairy, dark green leaves provide an interesting texture in the landscape as well. Purple coneflower will grow 2 to 3 feet (0.6 to 0.9 m) tall with a similar spread. This plant is loved by bees and butterflies that jump from flower to flower to collect its nectar. Even small birds, such as goldfinches, will gravitate to this plant to feed on the seed of the cone-like flowers.

Landscape Value

Purple coneflower is excellent when mixed with other summer perennials in a mixed border or in a natural area with other grasses and wildflowers. Coneflower is considered pest resistant, making it desirable in gardens that prefer low-maintenance practices.

Cultural Requirements

This carefree perennial prefers moist, well-drained soil and full sun but will tolerate a wide variety of soils and partial shade. Every few years, plants should be divided to keep them vigorous and productive. Purple coneflower is hardy to winter temperatures as low as −40°F (−40°C) and will also take heat and humidity well.

Cultivars

'Kim's Knee High' is a shorter variety growing 12 to 24 inches (30.5 to 61 cm) tall and displaying pink flowers. 'Magnus' offers larger rosy purple flowers and seedheads. 'White Swan' is a beautiful white coneflower with pure white petals that are very striking when combined with traditionally colored purple coneflowers. Two other excellent series are the PowWow® Echinacea, which are relatively compact and floriferous, and the 'Cheyenne Spirit' mix, which offer a wide variety of colors.

Joe-Pye Weed (*Eupatorium* spp.)

Ornamental Value

There are several species and cultivated varieties that are suitable for the garden setting. Joe-pye weed ranges in size from over 7 feet (2.1 m) tall down to under 3 feet (0.9 m). The bluish purple, dense, flat-topped flower clusters blooming in late summer and fall and the dark green leaves make Joe-pye weed hard to miss. Bees and butterflies will benefit from the profuse late-blooming flowers as many summer flowers begin to fade. Joe-pye weed tends not to be a favorite food of deer.

Landscape Value

Joe-pye weed can be used in mixed herbaceous borders, and tall varieties can be used in the back of the border to add height to the landscape. Because it prefers moist soils, so Joe-pye weed can also be used in a rain garden or near the downspout of the house.

Cultural Requirements

Joe-pye weed needs ample moisture and full sun to perform well. This plant should not dry out, and organic soils are best to allow them to reach maximum size. Mature plants can get leggy, so cutting them back a few times earlier in the season will force plants to become bushy. It is hardy to temperatures as low as −30°F (−34.4°C) .

Related Species and Cultivars

A common species of Joe-pye weed, *Eupatorium purpureum*, is a robust grower to 7 feet (2.1 m) or more tall. Give it room; otherwise, you will quickly regret how much room it has taken up in your garden. Its flowers are purple, and this rather upright grower will grow in a wide variety of climates and can tolerate winter temperatures as low as −30°F (−34.4°C). *Eupatorium maculatum* 'Gateway,' at 5 feet (1.5 m) tall, is more compact and bushy with reddish purple flowers. One of the smaller garden varieties is *Eupatorium dubium* 'Baby Joe.' It offers beautiful lavender flowers and grows only 2 to 2½ feet (0.6 to 0.8 m) tall, making it ideal among lower-growing perennials.

Cranesbill (*Geranium* spp.)

Ornamental Value

These perennial geraniums should not be confused with the annual bedding plant type known as zonal geraniums (*Pelargonium* spp.) that have white, pink, or bright red flowers. The perennial geraniums, called cranesbill because their fruits look like cranes' beaks, are generally easy to grow and adaptable in the landscape. They range in size and color from a 6-inch (15.2 cm) low, mounded groundcover to a bushy plant 4 feet (1.2 m) tall displaying a palette of colors that include white, pink, purple, magenta, violet-blue, and red. Perennial geraniums have foliage that is just as interesting as their flowers—often deeply cut and lacy. Perennial geraniums are not considered a regular favorite of deer but are frequented by butterflies.

Landscape Value

Cranesbill can be used in mixed borders and natural areas such as a woodland garden. Even when not in flower, the fine foliage and interesting growth habit make this plant desirable in the landscape. These reliable perennials will gently weave their way into the garden among other favorite plants with similar cultural requirements.

Cultural Requirements

In general, cranesbill thrives in moist soil and full sun or partial shade. Some species are rather shade tolerant, and in warmer climates it is wise to position your plants on an eastern exposure, giving them the cool afternoon shade they desire. Most cranesbills are hardy to −40°F (−40°C).

Cultivars and Related Species

Grayleaf cranesbill (*Geranium cinereum*) is hardy to −20°F (−28.9°C) and features showy pale purplish pink flowers with pronounced, dark purple veins. Bigroot geranium (*Geranium macrorrhizum*) is a European species growing to 18 inches (45.7 cm) tall with a similar spread. It has finely dissected leaves that, when crushed, have an aromatic fragrance. The plant offers purple-magenta flowers in spring and early summer. It has a spreading root system and will expand in size easily with the proper conditions. It is hardy to −40°F (−40°C). A hybrid geranium, 'Rozanne,' is an exceptionally good cultivar with violet-blue flowers that have white centers blooming in spring and summer and growing between 12 inches (30.5 cm) and 18 inches (45.7 cm) tall. 'Rozanne' is hardy to −20°F (−28.9°C).

Bee Balm (*Monarda didyma*)

Ornamental Value

This North American native plant displays whorled, tight clusters of bright red flowers in summer. The flowers are tubular in shape, making them ideal for bees, butterflies, and hummingbirds. Bee balm grows between 2 feet (0.6 m) and 4 feet (1.2 m) tall with upright stems and pointed, aromatic leaves that tend to turn deer off.

Landscape Value

Bee balm is typically used in a mixed border with perennials of similar height and texture. It is also effective in a sunny location near a stream or pond. One ideal use is planting bee balm in a rain garden, where regular moisture is available.

Cultural Requirements

Full sun or partial shade is best for bee balm, and soil must be moist. Rich, organic garden soil is ideal for this attractive garden favorite. Soil that dries out will spell doom for your bee balm. Stressed plants are typically more susceptible to foliar diseases such as powdery mildew. Every few years plants should be divided, as large clumps tend to die out in the center. Deadheading the flowers once they are faded will stimulate new flowers to form. Bee balm is hardy to −30°F (−34.4°C) and tolerates warmer climates as well.

Cultivars

New cultivars tend to be more resistant to disease. 'Fireball' has ruby-red flowers and semicompact plants growing about half the size as *Monarda didyma*. 'Marshall's Delight' is another variety that has been rated as good for disease resistance. It provides purplish pink flowers in summer and grows about 3 to 4 feet (0.9 to 1.2 m) tall.

Catmint (*Nepeta × faassenii*)

Ornamental Value

Catmint is a vigorous perennial with gray-green foliage and lavender-blue flowers much of the summer. The dense, mounded habit will reach 18 inches (45.7 cm) tall rather quickly and spread into dense clumps. The aromatic leaves will be cherished by your cat, who will roll in it; but deer will leave this plant alone. Bees and butterflies will also be frequent visitors to these hardy summer bloomers.

Landscape Value

Catmint is a rugged, durable, and adaptable plant working pretty much wherever you use it. It is ideal in mixed perennial borders and groupings and along walkways, rock gardens, and herb gardens. It can be used in a similar way to lavender but will grow where no lavender dares to.

Cultural Requirements

Catmint prefers moist, well-drained soil and full sun or partial shade. It will also adapt quite well to drier, rocky soils and hot, dry locations; but it benefits from afternoon shade in warmer climates. Cutting dead flowers will often encourage reblooming, and dividing large clumps at the end of the season will help keep plants vigorous. Catmint is hardy to −40°F (−40°C).

Cultivars

'Six Hills Giant' is a robust grower to 3 feet (0.9 m) tall with showy violet-blue flowers that dazzle the summer garden. 'Walker's Low' should not fool you, because its name does not refer to its size. This vigorous grower can reach 18 inches (45.7 cm) tall and a slightly wider spread. It is an outstanding performer and will rebloom late into the season if cut back and properly cared for.

Black-Eyed Susan or Goldsturm Coneflower (*Rudbeckia fulgida* 'Goldsturm')

Ornamental Value

The Goldsturm coneflower, sometimes called black-eyed Susan, blooms profusely in summer and fall with bright, golden yellow flower petals surrounding dark brown to black centers. The 18- to 24-inch-(45.7 to 61 cm) tall plants with thick, dark green leaves form large masses within a few years. This plant will serve as a magnet for bees and butterflies.

Landscape Value

Goldsturm coneflower is a showy, colorful plant that is quick to establish as a standalone mass planting or in groups with companion plants such as purple coneflower, ornamental grasses, and summer-flowering shrubs. Not much can beat it for late-season color in the garden.

Cultural Requirements

Goldsturm coneflower is a particularly vigorous variety with few demands. It thrives in well-drained, moist soil but is remarkably adaptable to a wide range of soils. It will also perform well in partial shade. It is drought tolerant and is not prone to pest problems. Occasional dividing in the fall will keep large clumps in check and of a reasonable size. This plant is hardy to –40°F (–40°C).

Related Species

Rudbeckia hirta 'Herbstonne' ('Autumn Sun') is a big, bold cultivated variety with long, drooping sulfur-yellow flowers that have green centers and a tall, upright growth habit to 5 feet (1.5 m) tall. It is hardy to –30°F (–34.4°C).

Autumn Joy Stonecrop (*Sedum* × 'Autumn Joy')

Ornamental Value

This tried-and-true succulent perennial is one of the best at attracting bees, butterflies, and gardeners alike because of its vivid flat-topped pink flowers that age to a rich reddish bronze. Autumn Joy stonecrop flowers from late summer into the fall, and the seedheads will persist through the winter. The fleshy, thick green leaves will emerge from the base in spring, eventually forming dense clumps of foliage reaching 12 to 24 inches (30.5 to 61 cm) in height.

Landscape Value

Autumn Joy stonecrop is excellent when mixed with other late-season bloomers such as asters, goldenrod, and ornamental grasses. There is nothing quite like a mass planting of stonecrop fluttering with the activity of butterflies in the late summer.

Cultural Requirements

This plant prefers well-drained soil and full sun. It is adaptable, but too much shade will cause weak, unhappy plants that will topple over. If sited in partial shade, cutting back your plants in early summer will keep them dense and less likely to flop. This may also cause smaller but more numerous flowers. Clumps can be divided in spring and will also keep mature plants from becoming too large. It is hardy to −40°F (−40°C).

Goldenrod (*Solidago* spp.)

Ornamental Value

Goldenrod is often maligned as a problem plant. It is accused of being invasive and causing hay fever (other species actually trigger hay fever) for people who have pollen allergies. But true goldenrods are excellent wildflowers and are important to wildlife such as bees, butterflies, and birds. The feathery, bright golden yellow flowers are unmistakable in the late summer and autumn garden. *Solidago canadensis* is found in prairies and native meadows across North America. Frankly, unless they are already growing in a nearby meadow or natural area, I would recommend sticking with a few good garden varieties of goldenrod. Goldenrods are also known to be reliably deer resistant.

Landscape Value

Goldenrods are showy, upright perennials with soft plumes of flowers, making them ideal companions for late-season plants such as ornamental grasses, black-eyed Susan, and stonecrop in a mixed border. Goldenrod can also be used in a wildflower meadow or among native grasses in an open field. They are often admired from afar along roadsides.

Cultural Requirements

Rich, organic soil and full sun are best for goldenrods; and well-drained, loamy garden soil works just fine.

Cultivars

Solidago rugosa 'Fireworks' is an outstanding cultivar growing to 3 feet (0.9 m) tall with a dense, upright habit. The dense clusters of blazing yellow flowers look like streams of fireworks cascading off the plant. This variety is hardy to –30°F (–34.4°C).

Solidago sphacelata 'Golden Fleece' offers tight plumes of bright yellow flowers and a fairly compact growth habit 18 inches (45.7 cm) tall. This variety is also hardy to –30°F (–34.4°C). *Solidago shortii* 'Solar Cascade' is similar in size to 'Fireworks,' with beautiful arching branches loaded with showy yellow flowers; but it is a bit hardier to –40°F (–40°C).

Aster (*Symphyotrichum* spp.)

Ornamental Value

Asters are related to chrysanthemums and offer showy, daisy-like flowers ranging from white to pink to blue-purple in summer and fall. They typically grow several feet tall with delicate, dark green foliage. For late-season interest, asters are one of the most colorful and floriferous, attracting butterflies, bees, and other pollinators.

Landscape Value

Asters are good in mixed borders with other mid- to late-season perennials and can also be used in groupings or mass plantings. Some species and varieties often need staking in the garden to keep them from flopping.

Cultural Requirements

Asters in general prefer moist, well-drained soil and full sun, although partial shade is also acceptable. Cutting asters down to the ground while they're dormant in winter or early spring will help develop dense and productive plants. The asters mentioned below are hardy to –30°F (–34.4°C).

Cultivars and Related Species

There are many species of asters, but here are a few common ones that will make a good home in your garden. Frikart's aster (*Aster × frikartii*), a garden-friendly hybrid, will get 2 to 3 feet (0.6 to 0.9 m) tall by midsummer, flowering through the fall with its bright lavender-blue flowers. New England aster (*Symphyotrichum novae-angliae*) is a native aster found along the eastern United States. It grows 4 to 6 feet (1.2 to 1.8 m) tall with beautiful violet-purple flowers that have bright yellow centers blooming in late summer and fall. New York aster (*Symphyotrichum novi-belgii*) is a common roadside weed with violet flowers growing 3 to 6 feet (0.9 to 1.8 m) high. There are several dwarf cultivars that range from 12 inches (30.5 cm) to 3 feet (0.9 m) tall, including 'Professor Kippenburg,' which grows only to 12 inches (30.5 cm) tall with lavender-blue, semidouble flowers, and 'Ernest Ballard,' which grows to 3 feet (0.9 m) with semidouble reddish pink flowers.

ANNUALS

A few good low-maintenance annuals to add color and attract pollinators to the garden are:

- Beggarticks (*Bidens*)
- Cosmos (*Cosmos*)
- Dahlias (*Dahlia*)
- Heliotrope (*Heliotropium*)
- Hummingbird mint (*Agastache*)
- Lantana (*Lantana*)
- Marigolds (*Tagetes*)
- Million bells (*Calibrachoa*)
- Tobacco plant (*Nicotiana*)
- Salvia (*Salvia*)
- Verbena (*Verbena*)
- Zinnias (*Zinnia*)

THINKING OUTSIDE THE BOX:
Using Ornamental Grasses in a Cultivated Garden

Ornamental grasses have been a part of the cultivated garden for many years. Ornamental grasses began their meteoric rise in popularity in the 1980s when Wolfgang Oehme, a German-born landscape architect, and American James Van Sweden began to incorporate ornamental grasses into their landscape design projects. Piet Oudolf, another world-renowned garden designer, has also brought the idea of naturalistic landscapes to the forefront with his spectacular work.

For the past three decades, ornamental grasses have become a staple in well-designed perennial gardens around the world. Ornamental grasses provide professional garden designers with a soft, flowing texture and natural look as well as versatility and adaptability in the landscape. Home gardeners love ornamental grasses because they are relatively easy to grow and complement other flowering plants in the landscape.

But some of these ornamental grasses—such as Japanese blood grass (*Imperata cylindrica*), maiden grass (*Miscanthus sinensis*), and fountain grass (*Pennisetum alopecuroides*)—have become invasive in some areas. Luckily, over the past few years, there has been a real push by ecologists, horticulturists, and nursery professionals to introduce and breed more noninvasive grasses for inclusion in the cultivated landscape.

Here are a few of my favorite ornamental grasses. What makes these grasses desirable is their clumping growth habit, meaning they won't spread very far in the landscape.

ORNAMENTAL GRASSES

Appalachian Sedge (*Carex appalachica*)

Ornamental Value

This beautifully fine-textured native sedge forms tight clumps to 12 inches (30.5 cm) tall with a graceful, flowing habit. It blooms in early spring, producing small, somewhat inconspicuous flowers and seedheads.

Landscape Value

This low-maintenance sedge is ideal in groupings, as an edging plant, in containers, and even in dry shade. Sedges are also effective in no-mow lawns where mowing is very infrequent or not at all.

Cultural Requirements

Sedges in general are durable and adaptable to soil and light conditions. Appalachian sedge prefers moist, well-drained soil and partial shade but is quite adaptable to a wide range of soils, including dry conditions. Partial or full shade is tolerable to this plant. Morning sunlight and afternoon shade are recommended. This plant is hardy to −40°F (−40°C).

Cultivars

Ice dance sedge (*Carex morrowii* 'Ice Dance'), a noninvasive Japanese counterpart, offers variegated clumps of foliage to 12 inches (30.5 cm) tall. This plant is a carefree groundcover that will also work in a wide variety of landscape situations. It looks similar to the commonly used and sometimes overused liriope. Cut back damaged foliage in early spring to tidy up plants. This plant is hardy to −20°F (−28.9°C). Other good sedge species worth pursuing are *Carex pensylvanica* and *Carex buchananii;* both are good foliage plants in mixed borders, natural settings, or containers.

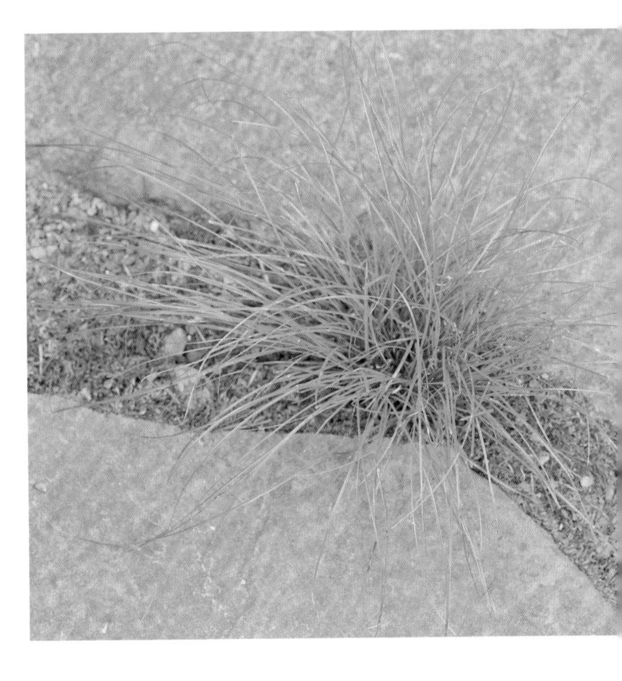

Big Bluestem (*Andropogon gerardii*)

Ornamental Value

Big bluestem is truly the king because of its large stature in the landscape. It exhibits silvery gray to blue-green foliage that turns to a coppery red in fall. The plant seems to glow in the winter landscape against a blanket of snow. The flower stalks turn a beautiful purplish color late in the growing season. Big bluestem can reach 4 to 6 feet (1.2 to 1.8 m) tall with about half the spread.

Landscape Value

This magnificent native is often found in meadows and prairies along roadsides. It can be used in natural areas and in groupings, where it will naturalize.

Cultural Requirements

Big bluestem prefers full sun and well-drained soil but is very tolerant of heavy clay soil and drought conditions. It is hardy to −30°F (−34.4°C).

Feather Reedgrass (*Calamagrostis × acutiflora* 'Karl Foerster')

Ornamental Value

Feather reedgrass is an upright, narrow-growing grass that ranges in height from 2 to 4 feet (0.6 to 1.2 m). Its foliage changes from green in summer to beautiful golden yellow in the fall. Pinkish flowers are borne on erect stalks in early summer and change to golden brown by late summer.

Landscape Value

Because of its upright habit, this grass can be used as a stand-alone plant or in groupings. It is effective in combination with other sustainable plants such as purple coneflower, black-eyed Susan, and Joe-pye weed. Feather reedgrass can also be used in rain gardens because it likes moisture.

Cultural Requirements

Feather reedgrass prefers full sun and moist soils that do not dry out. It is tolerant of clay soil and is hardy to −20°F (−28.9°C).

Little Bluestem (*Schizachyrium scoparium*)

Ornamental Value

Little bluestem is easily one of my favorite grasses. It forms upright clumps of blue-green foliage in summer with purplish bronze flower heads and clusters of fluffy, silvery white seedheads, which are striking. Plants will reach 2 to 3 feet (0.6 to 0.9 m) or more by the end of the summer. The foliage turns brilliant shades of golden yellow to reddish bronze and glows in the winter landscape. Huge masses of little bluestem resemble amber waves of grain as they sway in the wind.

Landscape Value

Little bluestem is most effective in groups and masses in natural, open areas of the garden. Some new varieties can be effective when used in combination with late-season perennials, but I prefer my little bluestem on its own in an open area of the garden.

Cultural Requirements

This tough, adaptable grass tolerates dry or moist soil and thrives in moist, well-drained soil and full sun. It is remarkably drought tolerant and requires very little to succeed. It does not need fertilizer and may flop late in the season in fertile, rich soils. In the early spring, cut plants down in preparation for a new batch of foliage. This is no doubt one of the most sustainable grasses out there because it has to be mowed only once a year and can survive in a wide variety of conditions. It is hardy to −40°F (−40°C) and can also thrive in warmer climates.

Cultivars

'The Blues' has distinct blue-green foliage changing to shades of reddish bronze in fall, while 'Blaze' is grown for its vivid red fall and winter color. 'Standing Ovation' is an upright form that is less likely to flop in a garden setting.

Pink Muhly Grass (*Muhlenbergia capillaris*)

Ornamental Value

Pink muhly grass looks like any other grass when not in bloom, growing between 1 foot (0.3 m) and 3 feet (0.9 m) in height. But in fall, big clouds of puffy pink flowers develop, creating an unbelievable show that is nothing short of stunning.

Landscape Value

Pink muhly grass is very effective in large masses or smaller groupings mixed with other fall-interest plants.

Cultural Requirements

Pink muhly grass prefers full sun or partial shade and thrives in soils with good drainage. It is rather heat and drought tolerant and thrives in these conditions. It will also tolerate windy sites. It is hardy to 0°F (−17.8°C) and is best suited for moderate climates and must be protected in colder climates.

Cultivars

'White Cloud' displays masses of white flowers in fall. While it is not as heart stopping as its pink relative, it is still a handsome grass for the landscape.

Prairie Dropseed (*Sporobolus heterolepis*)

Ornamental Value

This meadow grass can also serve as a specimen plant in the garden. It forms soft-textured mounds of foliage with open, airy flower clusters in fall, reaching 2 to 3 feet (0.6 to 0.9 m) tall. The flowers have a subtle fragrance similar to that of coriander. The foliage turns rich shades of golden yellow to orange in autumn.

Landscape Value

Prairie dropseed is excellent in small groupings or in large masses. It can also be used in open natural areas and in rain gardens.

Cultural Requirements

Prairie dropseed prefers dry to moderately moist, well-drained soils but adapts to many soil types. It is also quite heat and drought tolerant. It is hardy to −40°F (−40°C).

Switchgrass (*Panicum virgatum*)

Ornamental Value

Switchgrass forms dense, upright tufts of growth. The olive-green to blue-green foliage and profusion of loose, airy flower heads make this grass very desirable. Flower heads turn from pink to beige as they age. In the fall, plants will turn yellow and eventually tan for the winter. Switchgrass will eventually reach 3 to 6 feet (0.9 to 1.8 m) tall, making it a possible replacement for the more invasive maiden grass (*Miscanthus sinensis*).

Landscape Value

Switchgrass likes moisture and can be used in groupings or mass plantings by itself or with flowering herbaceous plants. It can be used in natural areas and rain gardens as well.

Cultural Requirements

Switchgrass adapts to a wide variety of soils, from dry to moist. It will thrive in moist, well-drained soil and full sun or partial shade. It is hardy to −30°F (−34.4°C) and can thrive in warmer climates as well.

Cultivars

'Dallas Blues' is a distinct variety of switchgrass with wide, blue foliage and large flower heads that are tinged with purple. 'Heavy Metal' emerges in the spring with a striking metallic blue color, which changes to a golden yellow in the fall. 'Northwind' is a distinctly vertical form with a compact habit to 4 to 5 feet (1.2 to 1.5 m) tall. The showy yellow flower heads will develop in late summer, making this fine-textured plant nearly 6 feet (1.8 m) tall by the end of the season. 'Shenandoah' offers reddish pink flower clusters and a beautiful burgundy-red fall color that is sure to brighten up the autumn garden. Mature plants with foliage and flower heads will reach 4 feet (1.2 m) tall.

Japanese Forest Grass or Hakone Grass (*Hakonechloa macra* 'Aureola')

Ornamental Value

Variegated Japanese forest grass is a rather well-behaved species with a spreading, graceful habit and two-tone green-and-yellow leaves. Only growing 12 to 18 inches (30.5 to 45.7 cm) tall and wide, this grass turns a beautiful yellow-bronze color in fall.

Landscape Value

Japanese forest grass makes an excellent groundcover or edging plant and can be mixed with other ornamental grasses and perennials. It is especially effective as an accent plant in a shade garden. It will brighten up any dark spot in the garden.

Cultural Requirements

This plant does best in moist, well-drained soil and partial shade. It will tolerate full shade but will often lose some of its bright and variegated color. Japanese forest grass should not be allowed to dry out, as it will not perform well. It is hardy to −20°F (−28.9°C).

Cultivars

'All Gold' is a brightly colored variety with leaves that are completely golden yellow.

Mexican Feather Grass (*Nassella tenuissima, Stipa tenuissima*)

Ornamental Value

This beautiful grass has fine-textured, feathery foliage forming clumps and growing up to 2 feet (0.6 m) tall. The fine green foliage is covered by silky, soft flower heads that form in early to midsummer before they change to a striking light golden blond color. It is a sight to behold when this plant gently sways in the summer wind.

Landscape Value

Mexican feather grass is quite striking in masses along a walkway or edge of a path. I have seen it used very effectively with tall species of verbena (*Verbena bonariensis*) and in combination with other ornamental grasses.

Cultural Requirements

Mexican feather grass prefers moist, well-drained soil and full sun. It is fairly drought resistant once established. Cut back plants in early spring to remove old foliage. This plant is hardy to 0°F (–17.8°C) and is best suited for warm climates.

CREATING SUSTAINABLE GRASS MEADOWS: NO-MOW ALTERNATIVES TO THE SAME OLD LAWN

For some time now, I have been observing and admiring grass meadows. I continue to be amazed by their complexity and sheer beauty when they are done right. Of course, I think nature creates grass meadows better than humans ever could, but if you have the time and inclination, why not try? Grass meadows are among the most eco-friendly and sustainable ecosystems in the landscape. They support a delicate and important relationship between plant life, wildlife, and soil organisms that all depend on one another. Our role is to preserve and protect these ecosystems—and sit back and enjoy them!

I used to think that creating meadows was easy: Just sprinkle some seed on an open area or stop mowing a grass field and—voilà!—instant grass meadow. This, however, is far from the truth. It takes time and effort to create grass meadows right. Once the initial planning and preparation are done and the meadow is established, then you can sit back and admire them. Unfortunately, all too often, outside factors such as invasive species can cause challenges when establishing a grass meadow. But grass meadows are gaining in popularity, and in some cases they are replacing the traditional manicured lawns found in so many landscapes.

Letting a field, pasture area, or part of the lawn grow long for the summer months is very eco-friendly.

Native meadows are beautiful and sustain a wide variety of wildlife.

Grass meadows offer many benefits. Over time, grass meadows can support a diversity of plants rather than the monoculture that exists in a traditional mowed lawn. The combination of grass species and wildflowers can add interest most of the year. In addition, grass meadows will attract and sustain a wide variety of birds, mammals, and insects that benefit the garden. Although grass meadows require a lot of planning and preparation, once established, they are far easier to take care of than mowed grass. Meadows do not require regular watering or fertilizer and get mowed only once a year as opposed to once a week. After a grass meadow is established, regular monitoring to keep out invasive weeds is required. But think of how much time and maintenance you will save, not to mention fuel for your mower. Is week after week of backbreaking work to keep your lawn mowed, edged, and weed-free really worth it? There is more to be discussed about this issue, but if you decide to plant a grass meadow, it will probably need to be in the backyard or in areas along the fringe of the property. Many communities have specific laws prohibiting grass meadows or "weedy" plantings in the front yard and up to the sidewalk. Don't worry, though; there are many solutions to this problem.

Site Selection and Preparation

My best advice if you're planning to convert your lawn or garden to a grass meadow is to start small. Select a 5 × 5 foot (1.5 × 1.5 m) area of the garden and use it as a test plot. This will ensure you truly like the meadow concept and work out any kinks before you invest in a bigger section of the garden. Choose a site that is in full sun or at least partial shade with a good amount of light. Dig in the soil and determine what type of soil and drainage you have. This will help you determine what types of meadow grass species will grow on your site. Choose the types of grasses that you want to grow and try to find a reliable source for seed or small plants. There are two categories of grasses: cool-season and warm-season. Cool-season grasses grow in the spring and early summer, when the weather is cooler, while warm-season grasses thrive during the summer, when the air and soil temperatures are warm. It is wise to have a combination of both in a meadow planting, provided they will not compete with one another.

The second step in creating a sustainable grass meadow is to get rid of your lawn or undesirable vegetation growing in the area where you want your meadow to be. This can be done by cutting out the grass or plants with a shovel into smaller sod strips. Another way to get rid of unwanted vegetation is to cover it for a few months with black plastic or landscape fabric. I have also seen gardeners use a generous layer of newspaper covered with a thick layer of wood chips to hold it down. These methods could be used in the fall or very early spring to ensure that the plot of land you are using is free of vegetation—most importantly, weeds. Although nonselective herbicides can be used in spring to clear an area effectively, why not use nonchemical methods? I would recommend using herbicides only if the area you are planning to convert to a grass meadow is very large. Whatever method you choose, your site must be smooth and free of weeds before the meadow is planted.

Planting Your Meadow

Spring is an excellent time of year to plant your grass meadow, especially for warm-season grasses. Fall planting is acceptable for cool-season grasses, and often a cover crop helps protect the bare soil. Rototill or plow under the soil and smooth it out with a garden rake. Remove any large clumps of dead vegetation or roots, large rocks, and so on before planting. Seed can be spread by hand or with a broadcast spreader. You should read the directions carefully when sowing your grass seed. After sowing, apply a light layer of sifted compost or fine mulch and water the seed well. Native meadow grass germination and establishment can take some time, so be patient. If your area is small enough and you choose to plant small plants instead of seed, establishment of your meadow will be quicker and controlling invasives a bit easier.

After your new grass meadow has germinated and started growing, it is important to patrol it regularly for invasive weeds. The first few years are most critical to your new and vulnerable meadow. Spot weeding by hand and eliminating weeds before they go to seed is important. Spot herbicide applications can be done on noxious, hard-to-eradicate weeds. The first year, you can mow your meadow to about 4 inches (10.2 cm) high to reduce the likelihood of weed plants going to seed. In the second year, you can reduce mowing to once a year in the late fall or early spring. I prefer mowing in spring to give birds and other wildlife winter cover.

ONE GARDEN AT A TIME

While data point out that the average mean temperature across the globe has risen over the past few decades, the debate over whether this is a short-term phenomenon or will have longer-lasting consequences continues. Whether you agree or disagree with the idea that human actions are contributing to climate change, there is no denying the fact that sustainability is crucial to our future. By putting the right plant in the right place and following sound gardening principles, we can help the environment in which we live one garden at a time. My hope and belief is that over time, the global warming trend will reverse course and weather will return to historic patterns. Regardless, climate change is affecting where and how we grow plants today.

CHAPTER 3

MANAGING UNWANTED GARDEN PESTS WHILE NURTURING WILDLIFE

Managing the Bad Guys while Encouraging the Good Guys

Controlling damaging pests can be one of the most challenging aspects of gardening. Doing so while limiting the amount and frequency of pesticides that are applied to the garden can be even more of a challenge. Worldwide there is an environmental movement to reduce the overuse and misuse of pesticides and chemical fertilizers in an effort to reduce pollution, protect valuable water supplies and native habitats, and reduce many health hazards linked to irresponsible synthetic chemical use. The idea of using sound gardening practices and plant health care strategies to effectively and efficiently control harmful pests while reducing our dependency on pesticides and other chemicals is a major component of the sustainability movement. The conventional practices of using pesticides without regard for the environment have spurred criticism from political leaders, environmentalists, and health officials. In the past, it was common for home gardeners to apply pesticides regardless of whether a problem existed. It was not at all uncommon to see garden pesticides applied as *preventatives* in case pest problems arose or in response to a past problem that homeowners were afraid would recur. Too often our hearts are in the right place, but our minds are on autopilot about the use of pesticides, chemical fertilizers, and other potentially harmful products.

HOW TO IMPLEMENT A SOUND PEST MANAGEMENT PLAN

Sustainable pest control is an approach that utilizes regular monitoring to determine if and when treatments are needed. It incorporates various forms of pest control, such as physical or mechanical removal of pests, cultural practices, crop rotation, biological controls, and even plant selection to keep pest numbers low enough to prevent intolerable damage or annoyance. It uses information about plants, pests, and the environment to plan and implement effective management strategies. The first line of defense is to manage your landscape and address pest issues before they become a problem.

Managing your garden sustainably is not about just seeing a pest and spraying it with a pesticide. It is more about taking responsible steps to first evaluate if the problem even needs to be controlled in the first place. The use of chemical controls such as pesticides is usually considered a last resort. If a pesticide is needed, often a gardener will use the least toxic pesticide possible to do the job. By integrating many forms of pest control rather than just one, you can effectively and efficiently control pests while reducing the risks of pesticide exposure.

The key to a successful pest management program is monitoring pest populations to aid in the decision-making process. The overall intent of this process is to identify and reduce factors causing pest problems in the first place. This means getting to the root of the problem and addressing it before it becomes an issue. The true essence of a sound pest management program is to choose a treatment method based on your observations and then implement this treatment in the manner least disruptive to the natural environment.

Not all bugs you find in the garden are pests that need to be controlled. Many, such as this ladybug, are helpful to the gardener.

Japanese beetles are one of many garden pests. They feed on leaves of roses and other ornamental plants.

The fundamental philosophy of sustainable pest management is to integrate strategies to control or eliminate pests by first relying on alternatives to pesticides. This often results in the reduction of pesticide use because the resourceful gardener has found another way to manage the pest problem. While pesticides are one of several pest management tools that is available, they are usually considered a last resort.

Pests can cause various degrees of harm to plants, including economic and aesthetic ones, and health risks in humans. For example, high pest populations can cause crop failure at a farm, which can have a serious economic impact by causing rising prices at the grocery store. Pest damage can also significantly reduce a homeowner's property value by damaging valuable landscape plantings. Finally, pests can cause medical problems, such as West Nile virus (spread by mosquitoes) or Lyme disease (caused by ticks). So, when implemented correctly, these efforts have great value beyond just helping you maintain a beautiful garden.

MANAGING PEST POPULATIONS

Within the realm of insects and plants, there are good guys and bad guys; and in nature, there is a delicate balance between the two. In the garden setting, with a constant influx of new and exciting plant species and varieties being introduced, keeping this balance is even more difficult. Invasive insects and plants have become a serious issue worldwide, and the environmental and economic impacts are significant.

Monitoring is a key element of any successful pest management program. Regularly inspecting the landscape is critical because it will help identify potential problems before they become an issue. A gardener must know what is going on in his or her own backyard to properly manage it. The occurrence of pests on landscape plants varies from year to year and in different locations within the garden. But at the same time, patterns or cycles may form, allowing you to be better prepared the next time a similar issue arises. In addition, monitoring allows gardeners to anticipate proper timing of control methods. Monitoring the landscape and regularly observing pest levels allow well-informed gardeners to make responsible decisions about pest management. In many cases, pesticides may not be the best option; or in some cases, no action may be needed at all.

In conjunction with monitoring procedures, setting thresholds is also desirable. Thresholds, also known as tolerance levels, are the point at which gardeners may need to take some form of action to protect their plants. If a pest problem is bad enough and is now at the point where it will cause significant harm to your plants, it's time to implement some form of control. Every property has a different set of thresholds with no one site being the same as the next. Even within one site, tolerance levels may be different in certain areas. Perhaps less damage is tolerated in the vegetable garden than in a flower bed. A threshold is simply how much pest damage you are willing to tolerate before taking action.

Regular monitoring of your garden for insect and disease problems will reduce the chances of a serious infestation.

COMPONENTS OF A SUCCESSFUL PEST MANAGEMENT PROGRAM

Remember that a successful pest management program integrates many different forms of pest control measures to create an efficient, effective, and safe pest management plan. These measures include biological, biorational, physical, mechanical, cultural practices, plant selection, and occasionally chemical controls. One of these practices alone would be less effective, but collectively they are much more effective. If done correctly, this process will inherently reduce the need for pesticides. Here's more about each of these pest management strategies.

Biological Controls

Biological controls have become very popular, since they encourage the use of beneficial organisms to control pests. This is a way for gardeners to wage biological warfare on harmful pests safely and sustainably. Planting a diverse garden, filled with many types of plants, enhances natural biological controls, as it encourages a healthy population of pest-eating beneficial insects. Another example of biological control is releasing ladybugs or other predatory insects into the landscape to take a bite out of damaging pest populations. Several companies raise beneficial insects for sale. These organisms can be purchased and released into the environment as needed.

A praying mantis waits patiently to capture its next meal.

Biorationals

Biorationals are a group of pest controls that are relatively nontoxic and ecologically safe. These control methods can be synthetic or naturally occurring. Low-toxicity pesticides, such as horticultural oils, soaps, Neem oil, Bacillus thuringiensis (*B.t.*), and milky spore disease, are all good examples of biorationals that are effective and break down rather quickly in the environment, leaving no lingering effects, unlike conventional pesticides. Beneficial insects and microorganisms are also considered effective biorational control methods.

Physical Controls

Physical controls provide a nontoxic means of pest control and are quite simple to implement. Harmful pests can be removed or killed by physically removing them from your plants. For example, pests such as scale can be wiped off plants very easily with a cloth and rubbing alcohol. Although this is very time consuming, it is also very safe. Whiteflies, aphids, and spider mites can be suppressed by regularly dousing the plants they are on with plenty of cold water.

Mechanical Controls

Mechanical control is similar to physical control, but it usually involves some type of machinery. The use of tillers or plows to manage weed crops in an agricultural field is one good example of mechanical control. Gardeners can effectively use small, motorized soil cultivators to control weeds in garden beds or vegetable gardens. For pests, mechanical controls can involve the use of traps and lures to capture pest insects, or screens and row covers to keep them off of plants.

Pest traps like this cucumber beetle trap offer a way to mechanically control pests without the use of pesticides.

Cultural Practices

Cultural practices refer to sanitation to reduce disease or insect problems. Cultural practices also relate to creating optimum environmental conditions and reducing plant stress. This, in turn, will reduce the likelihood of plants being vulnerable to pest problems. For example, pruning and destroying diseased limbs from a shrub to reduce the chance that diseases will spread to healthy plants is a type of cultural control. In vegetable gardening, crop rotation with unrelated species will prevent pests from establishing in a garden.

Plant Selection

Among the various kinds of pest control available, none are more important or effective than proper plant selection. Using the right plant in the right place and following sound gardening practices, such as planting procedures and watering, will go a long way to achieving success in the garden.

Chemical Controls

Chemical controls include using any pesticides or other synthetic materials that kill or have other negative effects on pests. Pesticides are one of many options a gardener has to control pests, but they should be considered only as a last resort should the above measures prove ineffective. The responsible use of pesticides when needed is still a viable option in cases of extreme pest infestations.

Using a motorized dethatcher on the lawn can help revive dead grass and maintain a healthy lawn. It's one example of a cultural practice that reduces pest pressure.

KEEPING ACCURATE RECORDS

Once you have evaluated your landscape and have collected information based on your observations, put this information into a usable form. Gardeners should be well organized and prepared to deal with ongoing pest problems in the garden. Sustainability does not reduce your risk of pest occurrences, but it does enable gardeners to be more prepared to address the problem when it happens. Keeping a garden journal or daily logbook is a great way to stay focused and organized. While pertinent information such as planting and harvest dates, composting schedules, and watering schedules are all important to note, pest activity is also important to record. The following chart shows an example of the type of information that can be recorded in a daily logbook. This information can be handwritten, or charts like the one below can be created.

Note that the information collected is pertinent to a gardener's ability to make responsible pest management decisions. Information such as plant type, identification of pest, date observed, and treatments made are all recorded. This information should be kept in a safe, accessible location in the toolshed or the house and referred to year after year. This will allow gardeners to keep accurate records of past history in their garden. Since pest infestations often run in cycles, gardeners can anticipate problems before they occur.

Pest Management Record Keeper

Plant Name	Location	Problem	Date Monitored	Treatment	Treatment Date	Comments
Rose	North side of house	Blackspot	6/12	Neem oil	6/18	Moderate infestation
N/A	Next to front door	Canadian thistle weed	7/3	Hand pull	7/5	Noxious weed removed manually
Zinnia	Next to pool	Powdery mildew	8/2	Fungicide	8/10	Monitoring effectiveness of treatment
N/A	Southeast corner of deck	Poison ivy	8/12	Post emergent herbicide	8/15	Weed controlled within 2 weeks

WEED MANAGEMENT

There are many effective and eco-friendly ways to control weeds as well. For example, in patios and cracks of walkways, pour boiling water on weeds to kill them rather than using weed killers. Acetic acid (vinegar) and clove oil or citrus oil are sold in formulations that are organic, nonselective weed killers. A weed torch is a device that is powered by a small propane tank and is effective in walkways, patios, and driveways to burn down weed growth; but be sure to take extra care when using this type of equipment. In addition, putting down a layer of newspapers with mulch over them is an effective way to keep weeds at bay around your perennials, annuals, and shrubs. These are just a few examples of environmentally safe and effective weed control measures.

Carefully removing weeds by hand is necessary around your favorite garden plants.

NURTURING THE GOOD GUYS

Plants and insects have a mutually beneficial relationship that is essential to their survival. Plants provide food and shelter for insects, birds, amphibians, and other animals. Animals return the favor to plants by serving as pollinators, seed distributors, and pest reducers. Adding plants that attract these beneficial animals into the garden will encourage a more sustainable, self-maintaining environment.

There are a host of plants that not only offer beautiful flowers and colorful fruit and foliage but also act as insect attractors. The simple fact is that the only way to attract and keep these good guys in the garden is to offer them a regular source of food, water, and shelter. These plants can be natives or non-invasive exotics and can easily be incorporated into a cultivated or more natural garden

Include a wide variety of flowering plants in your garden to encourage beneficial insects.

setting. In addition to incorporating into the garden plant species that attract beneficial insects and birds, it is important to understand what role these animal good guys play.

Helpful insects such as bees, butterflies, and lady beetles are near and dear to our hearts as they flutter around the garden. Honeybees, native to Europe, are important pollinators. They are social, nonaggressive, and resourceful insects that create large colonies in the landscape. Not only do they benefit the garden as pollinators, but honeybees also provide us with honey, beeswax, and royal jelly.

It is important for us, as gardeners, to understand that we are not alone in our quest to develop more sustainable ways of living. Conventional thinking would have us believe that humans are at the top of the food chain and all other living things are secondary. History has shown over and over that humans have taken from the Earth more than they have given back. But the sustainability movement has surely demonstrated that humans are only part of a bigger picture within our global ecosystem. Humans, animals, plants, and *all* living organisms need one another to survive.

Wildlife is a very important component of a sustainable garden. Sustainability is not only about reducing waste and not overburdening our resources; it is also about how animals play an important role in our environment. Frankly, sustainability only works if we realize that we cannot do it alone. Animals such as insects, frogs, lizards, snakes, birds, bats, foxes, and so forth complete the sustainability cycle. Animals are our protectors, pollinators, seed dispersers, and planters. They do a much better job at all those things than we can, so we shouldn't take them for granted.

Insects That Spread the Wealth

Beneficial insects come in many shapes, colors, and sizes and serve different functions in the garden. One of the main reasons to avoid using pesticides indiscriminately is because by using them, you are likely to kill off more beneficial insects than pests. That is why when pesticides are needed, they are used to target a specific pest. Butterflies and bees are particularly sensitive to insecticide applications. By reducing unneeded hazards and providing a safe and desirable habitat for wildlife, the sustainability cycle will thrive. Following are several groups of beneficial insects, their functions, and how to keep them coming back for more!

Butterflies

There is nothing quite like a warm, sunny day when the garden is busy with the fluttering of colorful butterflies. Butterflies not only come in a wide variety of colors and sizes, but they are among the best plant pollinators, along with bees. Pollinating your plants is very important to the survival of the landscape. Pollination allows plants to develop seeds and fruit; it is estimated that about one-third of the food you eat depends on pollination. Most important, pollination facilitates plant diversity.

Butterflies have four basic requirements in order to thrive in the garden. They need full sun at least part of the day, shelter from wind and heavy rain, a source of water, and food. In order for butterflies to survive in your garden, you need to provide plants that are a food source for butterfly larvae (caterpillars) as well as plants that provide nectar and pollen for adults. Because caterpillars do eat plants,

Be sure to have insect-friendly plants in bloom throughout your garden all season long.

you must be willing to allow some damage to host plants, and you must be willing to avoid using insecticides regularly, as they harm both caterpillars and adults. Accept some damage to ornamental plant hosts. To alleviate this issue, you can site plants that caterpillars feed on in a less visible area of the garden while placing plants that provide nectar and pollen for adult butterflies in a more prominent location.

In addition, having a healthy mix of both native and exotic plants in mass plantings that offer blooms from spring until fall will support healthy butterfly populations. The types of flowers that are attractive to butterflies are also important to consider. Flat-topped or clustered flowers will provide convenient landing platforms for butterflies in pursuit of nectar. Brightly colored flowers offering a wide range of colors such as white, pink, purple, yellow, orange, and red will do the job.

It is important for butterflies to bask in sunlight to regulate their body temperature. One way to encourage basking is by scattering flat stones or other landing surfaces throughout the garden, where butterflies can rest and absorb sun and heat. A water source is also important for butterflies, and using items you already have around the garden can easily provide this. Take a small saucer from a potted plant and put a few flat or round river rocks in it. Then fill the saucer with water and watch the butterflies flock to it to rest and have a drink. Gardeners can also provide a container of wet sand or pour water on bare soil to create a muddy puddle, where butterflies can obtain salts and minerals.

The Garden Is Abuzz

Bees are all too often considered pests in the garden and are lumped under the same category as hornets and wasps. This couldn't be further from the truth, as true bees like honeybees, bumblebees, or sweat bees are vital to plant life and the entire ecosystem. Although some bee species can sting, they usually do so only in defense, if threatened. Generally, bees are social but nonaggressive, often too busy at work in the garden to care much if we are present.

Various types of native and non-native bees are rather valuable pollinators, transporting pollen from flower to flower and fertilizing plants. This fertilization process, in turn, allows seed production and the reproduction and diversification of vegetables, flowers, trees, and shrubs.

Bees thrive in an environment where certain plants that grow in their habitat provide plentiful nectar. The plants and trees that are particularly attractive to bees are clovers, sunflowers, Queen Anne's lace, yarrow, Joe-pye weed, apple, blueberry, and cherry, as well as fruits and vegetables such as melons, squash, cucumbers, and tomatoes. That is just a sample of a far more extensive list of preferred plants for bees, so make sure your garden is diverse.

Various bee species nest in the ground or in empty tunnels and cavities and prefer undisturbed areas of the garden such as wetlands, untilled garden plots, or unmowed fields. This habitat will provide a source of food, water, and shelter for these resourceful insects. Below is a list of considerations for habitat, food, and shelter for bees.

- Introduce masses (usually groupings of five or more plants) of native plants with overlapping bloom times.
- Provide undisturbed soil areas, wet or muddy areas, unmowed areas, stone or sand piles, windbreaks, or logs in and around the cultivated garden for nesting and shelter.
- Nearby streams, ponds, or ditches can provide a reliable source of water.

Coneflowers are excellent attractors of pollinators.

LADY BEETLES TO THE RESCUE! WHAT'S CRAWLING AROUND YOUR GARDEN?

Lady beetles, also called ladybugs, are not all alike; and there are actually about five thousand species found worldwide. Lady beetles may make us feel warm and fuzzy, but if you are a soft-bodied insect such as an aphid, soft scale, or a mite, you don't stand a chance. Lady beetles are voracious eaters, quickly devouring everything in their path. It is estimated that one lady beetle can consume five thousand aphids in its life cycle. Both the adult and larval stage will feed on pests in your garden. The larval stage is spiny and alligator-like and will eventually transform into an adult beetle with spots. Some species of lady beetles will also feed on pollen and nectar from plants. Ladybugs are voracious eaters and will control several species of soft-bodied insects.

These tiny wrecking machines can provide excellent pest control in the garden, provided there is a food source and shelter to keep them happy. The key is to have a healthy balance so that the lady beetles will sustain themselves without the pest population getting too large and taking over. That is all done as part of your monitoring of the garden on a regular basis. Here are a few key ingredients to make lady beetles happy.

- Lady beetles live wherever their food sources are active but mainly on crops such as vegetables, grain crops, legumes, and grass fields and on trees and shrubs.
- Lady beetles will eat nectar and pollen, so planting flowering herbaceous plants is beneficial.
- Do not spray insecticides in the garden if ladybugs are present.

It is important to note that while beneficial in the garden, some species of lady beetles can be a nuisance, as they can collect around windows of buildings and other structures in cooler weather to keep warm. Some non-native species can also outcompete local native species for food.

THE PROS AND CONS OF GOING ORGANIC

Going organic has many pros and cons, but for many food gardeners, it's a must.

Creating a more sustainable garden supports the idea of living a healthy, lower-maintenance, and organic lifestyle. There are advantages and disadvantages to caring for your garden in an organic way. Here are a few pros and cons to consider as part of a sound sustainability program.

Pros

- Organic gardening protects our environment, including soil, water, air, and the health of our families and pets.
- Although debated, studies suggest organically grown food is richer in nutrients and antioxidants and is not contaminated with harmful chemicals, making it safer.
- Gardening organically will save money because you will not have to rely on expensive store-bought fertilizers and pesticides.
- Conventional garden maintenance techniques tend to work for a short period of time and often involve the use of chemicals that offer a quick fix, while organic techniques offer long-term care and enhancement of soil, water, and overall health of the garden.

Cons

- Maintaining your garden organically with less or no use of chemicals is more labor intensive and time consuming.
- Without the use of chemicals, lower yields may be experienced because of increased pest infestations and slow-acting organic fertilizers.

Besides the obvious and important value of gardening organically to save our planet, there are also economic benefits and costs associated with organic gardening. Less chemical dependency certainly saves money on expensive chemical fertilizers, weed killers, fungicides, insecticides, and other pest control products. But these organic methods also come at a cost. First, organically grown produce and ornamentals are typically more expensive to purchase, since they require more costly methods to grow. Because there is less reliance on chemicals, there is generally more crop loss and more culling out of damaged crops, so it takes more resources to produce the same amount of sellable crops in commercial organic gardening. Since organic farmers do not rely on chemicals, more labor is required for weeding and spreading of organic fertilizer like manures. As with agriculture, home gardens that are maintained organically that don't rely on pesticides tend to be more vulnerable to disease and insect damage.

Besides the monetary costs, another factor to consider when going organic is the availability of resources. Sustainability is all about maximizing the resources around you. A gardener who does this will no doubt create a more sustainable landscape that is less expensive to maintain. Common sense should prevail when making the most of your natural resources. If you live in an area where pine straw is in great quantity, then you should use that material as mulch in the garden. Why purchase wood chips or another form of processed mulch that is foreign to your neck of the woods when you have an abundant natural form of mulch right under your nose? Sustainability gives gardeners an opportunity to reuse, recycle, and maximize available resources. It takes a commitment of time and an investment of money to get the program up and running, but it pays dividends for years to come.

The wise use of resources offers an easy way to be more sustainable in your home and garden.

CHAPTER 4

MANAGING WATER PROPERLY

Making the Most of Every Drop

Water—and its role in our lives—is all too often taken for granted. We go about our lives day after day never really considering what would happen if we one day didn't have fresh, clean water at our fingertips. Let's face it, in many parts of the world, water is scarce; and if you live in a place where there is an abundance of water for drinking, bathing, recreation, and irrigating your lawns and gardens, consider yourself fortunate.

Since ancient times, when humans adopted agriculture as a way of life, water has been a vital component of society. Ancient civilizations knew that without a steady and reliable water supply, they could not survive. The ancient Romans and Greeks built aqueducts and cisterns to move and store water. In more recent history, there has been a steady increase in the demand for water. The main theory behind this rise in water consumption was that as the worldwide population increased, so did the need for water. Technological advances, improved equipment and regulations, and increased awareness of the need for water conservation have all resulted in more efficient use of water.

In the garden setting, reducing the demand for water and its responsible use is important, now more than ever. Irrigation systems are more efficient and effective than ever before. Inground irrigation systems make life a lot easier because all the piping and irrigation equipment is underground and always in place. Inground irrigation systems can be manual or automatic. The equipment is the same for both; the only difference is that automated systems have preprogrammed controllers that operate the system.

BRUTE.

Proper irrigation techniques are another very important part of watering your garden. An improperly watered garden full of stressed, poorly developed plants is more likely to fall victim to an infestation of pests or diseases. Following proper watering techniques is a huge part of creating a sustainable garden. Using common sense and following a few simple rules will go a long way toward reaching plant nirvana.

I see landscape after landscape either over- or underwatered. It pains me to see how much water, time, and money are wasted by homeowners who water improperly. One of the biggest culprits in this madness is the automatic irrigation system with an irrigation timer or controller. This is a device that can be programmed to water your garden automatically according to whatever information you input. I am not a big fan of automated systems because they are often programmed incorrectly and give you less control of watering. There are so many variables that help determine proper watering, such as sun and shade, soil drainage, temperature, and so on. An automatic irrigation system takes very little of this into consideration. Although manual irrigation systems are more time consuming, I prefer them because you have complete control and can water as needed with more attention to details such as the amount of natural rainfall your garden has received. Many automatic irrigation timers have manual overrides, so you can operate them in both modes. Of course, if your garden is truly sustainable and low maintenance, over time you should be able to rely less and less on your irrigation system.

WATERING BASICS

The basic rule when irrigating your garden is that watering should be less frequently and more deeply. For example, many residential irrigation systems will run 5 to 7 days a week for an average of 30 to 45 minutes each day. This is considered frequent, shallow watering, which wastes water and promotes poorly developed, shallow root systems. Watering in this manner, especially on a hot sunny day, will only moisten the surface of the soil and will not penetrate down to the roots of your plants. Much of the water will evaporate as soon as it hits the surface of the soil or grass. By contrast, an example of good watering practices would be watering 2 or 3 days a week as needed for an hour or more at a time. (I say "as needed" because it's important for you to check soil moisture regularly and water when plants need it, not take an autopilot approach to watering.) This will make the best use of your water because it will penetrate the soil and encourage the development of a deep root system. Of course, the exact amount of water your garden needs will vary depending on soil type, drainage, heat and humidity, light exposure, and most importantly, the types of plants you are watering.

Another important rule of thumb is to water early in the morning, avoiding mid-afternoon and evening watering whenever possible. Watering in the morning, when it is cooler, reduces evaporation rates. It also reduces the likelihood of widespread development of disease because the plants' foliage has a chance to dry off during the day, unlike plants that are watered in the evening. Watering in the heat of the day is not ideal, because much of the water will evaporate before being absorbed by the soil. Evening is the worst time to water because foliage stays wet for too long, promoting diseases. So, whether you are watering your lawn, flower beds or trees, these sound watering principles should be applied.

IRRIGATION EQUIPMENT: A WORLD OF POSSIBILITIES

There is a wide variety of irrigation equipment available to gardeners today. Now more than ever, irrigation equipment is efficient and effective. Today's systems require less water volume and pressure than they did in the past. These efficient methods of distributing water save both money and water, and they are relatively easy to operate and repair.

Strategic placement of drip irrigation emitters can lower your water use.

Drip Irrigation

Drip irrigation, or micro-irrigation, has been used since ancient times but was refined during the twentieth century. Today's drip irrigation systems are among the most efficient and effective ways to supply plants with water. Drip irrigation is a method by which small volumes of water slowly drip to plant roots through a system of pipes, tubing, valves, and emitters, which lie on or just below the surface of the soil. The system can be hooked up to a faucet or other water source and comes with filters, pressure control valves, and backflow preventers. Drip irrigation kits can be purchased and are fairly easy to install. This effective water distribution method is now a viable option for home gardeners as well. Drip irrigation can be used in flower beds, vegetable gardens, shrub borders, containers, and hanging baskets.

- **Pros:** durable, long-lasting, good on slopes, moderately priced
- **Cons:** assembly and regular maintenance of equipment required

Soaker Hoses

Soaker hoses are porous hoses that can be attached to a garden hose or faucet to evenly distribute water to the soil. There are no emitters or other equipment to worry about, and they are simply installed by laying them evenly spaced on the ground about 12 to 18 inches (30.5 to 45.7 cm) apart. Soaker hoses range in length from 50 to 100 feet (15.2 to 30.5 m) or more, but connecting hoses longer than 100 feet (30.5 m) may reduce effectiveness (unless you have excellent water pressure). The beauty of soaker hoses is that they can be placed on the surface of the soil and left exposed or hidden with a thin layer of mulch. Just be careful that you do not slice the soaker hose when digging in the garden. Soaker hoses have many applications, such as in vegetable gardens and flower beds and around trees and shrubs.

- **Pros:** inexpensive, easy to install and operate
- **Cons:** fragile as they age, can look unsightly unless hidden

Soaker hoses are an inexpensive way to distribute low volumes of water where you need it.

Rotary and Mister (Spray-Type) Sprinklers

New types of sprinklers like this impulse sprinkler can save water and provide more uniform coverage

With new technology, inground sprinkler heads have become much more efficient and effective than they were 20 or 30 years ago. There are several types of sprinkler heads that automatically pop up when activated.

The first type is known as a rotary sprinkler, and it sprays a stream of water to a desired width and length as it rotates. These are designed to use lower volumes of water while providing uniform watering. Rotary sprinkler heads are made primarily of lightweight plastic and/or stainless steel. Rotary type sprinklers can be recessed in the ground and pop up when the water is turned on, or for applications where taller perennials or shrubs are obstructing the stream of water, taller rotary sprinklers can be installed in beds.

Misters will also pop up or can be installed above ground to water taller plants, but they deliver a finer spray of water than rotary types do, and they do not rotate. They also do not typically spray water as far as rotary types do and are better for areas where shorter applications of water are needed. Misters are excellent in flower beds, along narrow strips of grass or ornamental plants, and in mixed borders with shrubs and trees.

- **Pros:** efficient, convenient, uniform watering
- **Cons:** can be expensive to install, must be maintained, and winterized in cold climates

Irrigation Bags

Irrigation bags, also known by the brand name Treegators, are plastic or rubber bags with weep holes at the bottom that are specifically designed to slowly water trees and shrubs in the landscape. They come in several styles and are wrapped around the base of the plant and then filled with water. Over the course of a few days, the water slowly leaks out, evenly watering the plant. Irrigation bags are especially effective with new transplants and in times of drought.

- **Pros:** inexpensive, easy to install and maintain
- **Cons:** need to be filled regularly to be effective, must be secured when empty so they don't blow away

PERMEABLE PAVERS: WHERE IS YOUR WATER GOING?

Permeable pavers are the next big innovation in the landscape world. They are paving blocks that can withstand vehicular and foot traffic and can be used in place of asphalt, cement, and other commonly used materials for driveways, walkways, and patios. Permeable pavers are installed similarly to brick or slate, but the material used under the pavers is various grades and sizes of gravel, which provides both support and drainage. When the pavers get wet from rain or irrigation, the water doesn't run off in every direction as it does with non-permeable surfaces. Instead, the water seeps through the joints of the pavers and penetrates the ground and the soil. This is a significant innovation, because in many landscapes, too much water runs away from the garden, down the driveway or patio, and into the street or into drains. This often creates erosion problems and is a total loss of water that could be used for your landscape. Permeable pavers allow water to penetrate the ground, reducing erosion and flooding, and allowing water impurities get filtered through the soil. This process greatly reduces pollutants getting into local water supplies and allows water to get to the water table more effectively. Permeable pavers come in a variety of colors and sizes and can be creatively designed to enhance the landscape. In large applications, such as driveways and large patios, it is best to have a mason or landscape professional install permeable pavers. In smaller applications, such as a walkway, small patio, or border for flower beds, home gardeners can install permeable pavers themselves.

HARVESTING RAINWATER

There are several ways to harvest rainwater and maximize the benefits of one of nature's most valuable resources. Collecting rainwater in rain barrels and cisterns is an excellent way to save and redistribute water wherever and whenever it's needed, but another great option is to create a rain garden.

Rain barrels are a great way to harvest and reuse rainwater, but check with local authorities as rain harvesting is prohibited or regulated in some areas of the world.

The installation of a rain garden takes careful planning but can be a fun and rewarding project.

Rain Gardens

Rain gardens are now common in both commercial and residential sites. This technique is a bit different from harvesting rainwater and collecting it in barrels or cisterns because rain gardens divert water back to underground water supplies rather than saving it for future irrigation needs. The concept behind rain gardening is quite simple and ingenious; let nature do the work. Rain gardens are low-lying areas that collect rainwater from roofs, walkways, driveways, and other impervious surfaces. These water collection areas are landscaped with plants that are adapted to regular or occasional flooding. These are typically plants that like "wet feet" but do not require standing water all the time.

The concept of a rain garden is that as water accumulates and seeps into the ground, the plants, roots, and soil will help filter out impurities in the water before it makes its way into an aquifer or other groundwater supply. Rain gardens are very beneficial because they reduce storm water runoff and erosion, reduce pollution, and replenish freshwater supplies. When properly installed, they are also very beautiful, lush plantings that can be attractive features in the landscape.

There are some basic requirements you must consider when developing a rain garden for your own landscape. It is not always as simple as selecting a site in your garden that tends to flood during rainstorms and planting some shrubs and perennials in that area. Believe it or not, rain gardens require drainage. Water that stands for too long—several days, for instance—is not a rain garden. It's a pool of stagnant water, which can attract mosquitoes and other unwanted pests. A properly functioning rain garden should drain rather quickly, certainly within a day. Rain garden sites can be areas that naturally collect water after a rainstorm, or you can help the process along by digging out an area that is convenient for you. Here are a few quick tips to follow when selecting and designing a rain garden.

- Select an area of the garden where water naturally flows or collects. Rainwater can be directed to this area by using drainpipes connected to your house's downspouts. Rain gardens should be placed at least 10 feet (3 m) from the house and away from the septic system.
- Dig out the area with a shovel and ensure that the subsoil is well drained. If needed, remove heavy soil and replace it with coarse sand or gravel. The sides of the rain garden should be gradually graded toward the middle. Rain gardens are shallow depressions and do not need to be any deeper than 6 to 12 inches (15.2 to 30.5 cm).
- On top of the well-drained subsoil layer, spread a layer of topsoil for growing plants.
- Before planting, let the area sit for a few weeks. Observe if the soil is collecting water and draining properly.
- Once you are satisfied with the function of your rain garden, plant groupings of ornamental plants that will grow in this environment, such as native grasses, ferns, and other plants that can tolerate varying levels of moisture.

Another factor that you should consider when creating a rain garden is the size needed to accommodate the runoff generated by a given area. A good formula to follow is that a rain garden should be at least one-sixth the size of the area draining into it. If your roof or patio is 20 × 30 feet (600 square feet), or 6.1 × 9.1 m (55.5 m^2), you divide that by six to get the proper size for your rain garden. That means that your rain garden needs to be at least 100 square feet (9.3 m^2) in size.

If you do not have adequate room for a full-sized rain garden, you can create a miniature rain garden by using a big planter or container filled with water-loving plants. Put the container at the base of your downspout and it will absorb and deflect the water from running all over the garden. If you choose this route, make sure your container is large enough and heavy enough to accommodate significant water flow. Small, undersized containers will fall over, and plants will become potbound too quickly to serve a long-term function. Clay, ceramic, or cement pots are excellent materials for a miniature rain garden.

Larger rainwater harvesting systems can be designed to collect the maximum amount of water possible.

BiG

CHAPTER 5

THE NOT-SO-BIG LAWN MOVEMENT

Doing Less and Still Having a Beautiful Yard

Lawns have been popular for many years, but that is slowly changing. Several decades ago, a conversation began about the challenges lawns create for our environment and what management techniques could be put in place to mitigate the ill effects of the traditional lawn. The main problem with lawns is that by nature, they are ornamental plant species (often not native) that require a great amount of care to maintain at a high level of quality. Lawns are really a monoculture of the same or similar species of grass or blended species with similar characteristics. This contrived planting is foreign to the natural landscape and therefore requires a lot of care, including regular watering, fertilizer to keep it green, and pesticides to keep them healthy.

It is impossible to criticize or lament the use—and sometimes abuse—of pesticides and fertilizers that make their way into groundwater supplies without discussing homeowners who can't live without a putting green–quality lawn. The blame cannot be placed solely on the shoulders of golf courses and farmers, who require such chemicals to maintain their unique crops and landscapes. Do homeowners really need to go to such great lengths to maintain beautiful lawns? Hardly. The average lawn is too big these days, dominating any other landscape or green space in the garden. It is time to look at reasonable alternatives to the traditional lawn. Gardeners would be better served by focusing their efforts on creating diverse, well-landscaped gardens that feature a wide variety of herbaceous and woody plants with four seasons of interest. At the very least, reducing the size of your lawn can create opportunities to develop and enhance new garden features with far more long-term benefits.

NATURAL LANDSCAPING: A LAWN ALTERNATIVE

Natural landscaping, or gardening with native plants, basically involves using all types of herbaceous plants, trees, shrubs, and so forth, that are indigenous to the area you live in to replace lawn areas. Native plants, unlike lawns, offer biodiversity and help filter impurities from water while protecting soil. Natural landscaping also provides much better habitat and food sources for wildlife, which lawns do not. Natural landscaping is, by definition, the complete opposite of a manicured lawn. I believe that this form of gardening in an informal context can also include noninvasive exotic species that provide food for birds, beneficial insects, and other wildlife. This type of natural landscaping, within any setbacks that may be required by law, will help to build healthy plant communities, wildlife habitat, and aesthetic plantings that offer multiple seasons of interest. A more naturalistic landscape design veers away from the traditional practice of a high-maintenance, time-consuming, and resource-dependent landscape toward a more self-sustaining, low-maintenance, and functional garden. Try to designate a portion of your yard and make that a more naturalistic landscape to support wildlife and create a lower-maintenance garden. This can be accomplished by leaving the less-visible areas of the garden more wild while maintaining the front yard so your landscape offers natural beauty but does not turn into an overgrown mess.

A diverse, well-balanced landscape with a variety of native species will help to create a sustainable landscape.

NEW MOWING METHODS FOR BETTER SUSTAINABILITY

A reel mower requires more labor but is better for the environment.

If you cannot take the leap to remove all your lawn or transform it into a natural planting, then another alternative is to explore less traditional, more eco-friendly mowing methods. These mowing practices are designed to reduce stress on your lawn, increase vigor and plant health, reduce noxious weed invasions, and reduce maintenance.

The reality is that a sustainable lawn is a healthy lawn. By practicing sound mowing, your lawn will require less water, be more lush and full, and be less prone to insect and disease infestations. The absolute worst thing to do to a lawn is to mow it too short. The growing point at the base of the grass plant is called a crown. If you cut your lawn too short, the crown could be damaged, resulting in poor growth or death of your lawn. Lawns that are mowed too short are more stressed and more prone to pests, and they dry out a lot quicker. The more foliage you leave on your lawn, the longer the soil will take to dry out. Also, lawns that are mowed too short can easily be overrun by invasive weeds. The recommended mowing tips that follow will save time, labor, chemicals, water, and money!

Lawns are typically mowed between 1 inch (2.5 cm) and 2½ inches (6.4 cm) in height, depending on the species. Most lawn species will tolerate this, but by no means is this an ideal situation. A much more sustainable way to manage your lawn is to raise the mowing deck of your mower to a height of approximately 31/2 to 4 inches (8.9 to 10.2 cm). Or just put your mower deck at the highest setting and that will suffice. At first this will be a bit of a shock to the gardener who is used to a putting green–type lawn—and the lawn may look a bit shaggy just before it is cut—but this will ultimately reduce maintenance and save money. Lawns that are left higher like this in the summer months will grow thicker, have deeper root systems, and will compete with weeds more effectively. The key to success with raising your mower height is to mow often so that you don't have too much of an accumulation of grass clippings, or use a mulching mower that cuts the clippings into tiny pieces and redistributes them over the lawn without leaving chunks of grass behind (more on mulching mowers in the next section). The one-third rule is a good guide for gardeners: It recommends that you should never remove more than one-third of your grass height at one time when mowing. It is also important to mention that you never should mow your grass during times of drought and heat. A drought-stressed lawn is extremely vulnerable, and by mowing it you are multiplying that stress tenfold. You should stop mowing until your lawn is adequately watered.

Types of Mulching Mowers

There are various types of mowers that can be purchased for the home garden. Either a walk-behind push mower or a ride-on tractor mower will do the job. These rotary-type mowers can range from 20 inches (50.8 cm) wide all the way up to 72 inches (182.9 cm) or more, depending on the type and model. These mowers can be designed to catch grass clippings or leave the clippings on the lawn. One excellent feature that gardeners have available to them today are mowers that have a special mulching blade. A mulching blade will chop up the grass clippings very finely and spread them evenly on the lawn. This type of blade is not expensive, and it simply replaces the conventional blade on most mowers provided that it fits the mower deck properly and can be properly installed and secured. Check with your local lawn care service professional when doing this. Mulching and dispersing grass clippings on your lawn is highly beneficial because it is a type of composting. As the clippings decompose,

Using a rotary mower without a bag but with a mulching blade is a great way to recycle grass clippings.

they add nitrogen and other nutrients back to the soil. This sustains your lawn quite well and reduces the need for chemical fertilizers. Studies show that lawns maintained this way need 30 percent less fertilizer over the course of a year. More mulching and composting of your grass clippings will save time and money and will also reduce the chance of chemical fertilizers leaching into the soil.

What Drives Your Mower?

Today's rotary mowers can be gas-powered, battery powered, or electric. While gas-powered mowers are still the most practical for large lawns, electric and battery-powered mowers can be quite suitable for a smaller residential parcel of lawn. With the advancement in new technology, battery powered mowers are becoming more effective, efficient, and affordable than ever before. These mowers are more powerful and maintain a charge. Since the goal is to reduce the size of your lawn anyway, there is no reason why mowers powered by clean energy cannot be effective.

Keeping Sharp

Regardless of the type of mower you decide to use, it is very important that you keep your mower blades sharp and in proper working order. Over time, blades can become dull and even bent if you occasionally hit rocks and other debris. Make sure your blades are sharp and in good condition at the start of the mowing season. Blades should be sharpened at least once during the growing season, but if you are mowing in rough terrain, sharpening your blades several times a season is recommended.

AN ECO-FRIENDLY LAWN

Believe it or not, lawns don't have to be a high-maintenance, water-guzzling, and wallet-siphoning entity. There are many innovative and creative ways to transform your lawn into an environmentally friendly, easier-to-maintain landscape feature. But managing an eco-friendly lawn does come with a price. It requires gardeners to compromise and give up traditional methods for new ones. The recommendations below do not require the use of chemicals but do require more hands-on attention from the homeowner.

Steps to Planting an Eco-friendly Lawn

- Check soil type and pH to make sure they are appropriate for growing a lawn. If your lawn is happy, it is more likely to compete successfully with weeds and pests and to survive hot, dry weather.
- Identify any bare areas in your lawn in early spring and reseed or replant those areas before noxious weeds take over. Once weeds like crabgrass have infiltrated your lawn, the only recourse is to pull them by hand.
- Fertilize several times a year with an organic fertilizer. There are many to choose from, and most have low amounts of nitrogen in them—but corn gluten is high in nitrogen. If you're using corn gluten, make sure it is applied after any seed has germinated, as corn gluten will inhibit germination.
- Mow high and leave the grass clippings. Also, during times of drought, you may decide to stop watering and mowing and let your lawn go dormant. In severe cases, you can water every few weeks just to sustain your lawn; once the weather cools down in late summer or fall, your lawn should come back.
- In high-traffic areas where soil gets compacted and lawn is sparse, think about replacing the lawn with stepping-stones, permeable pavers, or gravel.
- Practice sound watering techniques by watering deeply and infrequently. Only water the lawn when it needs it, not when the automated clock decides.

SHOULD YOU PLANT A LOW-MAINTENANCE FRONT LAWN?

I recommend that home gardeners move away front traditional lawn maintenance practices, such as mowing short, watering a lot, and pumping pounds of chemical nitrogen into the landscape; but having a grass meadow in your front yard is not usually practical. Leaving your front lawn unmowed for the summer will likely horrify your neighbors and may violate some local ordinances. But there is always a healthy compromise in most landscape situations, and in this case there may be a way to achieve more than one objective. I have seen resourceful gardeners around the world that treat their lawn like the rest of their garden rather than a separate entity. For example, in Australia, a water-wise gardener let his front yard go natural; and it was both beautiful and functional. This area became a combination of grasses, wildflowers, and strategically placed shrubs and trees that all worked together nicely. Water-wise gardens are now popping up all over the world, especially in areas with hot, arid climates.

You may want to consider planting lower-maintenance grasses, such as fine or tall fescue (*Festuca* spp.), in your front yard, occasionally cutting it at a high setting every few weeks or even less often. These cool-season grasses tolerate heat and drought as well as soil compaction. Zoysia grass (*Zoysia* spp.) is a warm-season grass that tolerates heat, drought, foot traffic, and even saltwater and pollution. In cooler climates, it will turn brown in the winter when it goes dormant but will turn green again when the weather gets warm again in spring. Both fescues and zoysia grass require a lot less maintenance than most other species of turf grass species.

A NO-MOW LAWN

Keeping grass taller, especially in times of drought, is better for your lawn.

A relatively new garden trend, no-mow lawns, are emerging in commercial and residential landscapes as another opportunity to create a more sustainable garden. No-mow lawns are virtually maintenance-free while still providing a green carpet in the landscape. No-mow lawns are typically mixtures of low-growing, clumping, or spreading plants that require little or no mowing, are drought and pest tolerant, and don't require large amounts of fertilizer to sustain them. They can provide solutions for moisture conservation and natural habitat while protecting soil and cooling the landscape. Typically a mixture of fine fescues (*Festuca* spp.) are used for this purpose, but depending on where you live, a variety of cool- or warm-season species can be used in no-mow lawns including zoysia grass (*Zoysia* spp.), clover, mondo grass (*Ophiopogon* spp.), sedges (*Carex* spp.) and other low-growing plants that fill this need.

The Lawnless Garden

It may be hard to believe, but yes, you can have a garden with no lawn if you choose. The lawnless garden is slowly becoming popular in many home gardens as we trend away from landscapes dominated by lawns. Many homeowners, especially those with a small enough lawn to convert easily, are choosing to replace their lawn with something else. I have seen natural stone pavers, gravel, mulch, and groundcovers all used in place of lawns. All these options can be more sustainable than lawns and just as functional if you prepare the area correctly and monitor it for invasive weeds, erosion, and other issues.

Now, we all know there are creative ways to use hardscape or softscape materials such as wood chips, gravel, and bluestone pavers and other products to replace your lawn. If you use your lawn for recreation, these products may not be the best option; but perhaps using some of the lower-maintenance grass species would be. These products are probably best in more of a passive-use area than an active-use area.

As part of the sustainability movement, now is the perfect time to offer some plant-related alternatives to a lawn. In areas where you are looking for an aesthetic yet functional feature and do not need an open area to play catch with the kids, groundcovers are a good option. There are many good, durable, and hardy groundcovers that can add the lush foliage you are looking for without all the mowing and such that goes with it. Once groundcovers are established, watering, fertilizing, and other lawn care activities should slow to a bare minimum, depending on what you select. The use of groundcovers will still require occasional weeding and blowing to remove debris, but it will still be a far cry from the horticultural marathon of lawn maintenance.

Russian arborvitae (*Microbiota decussata*) is a dense, fast-growing groundcover that can be used in shaded areas instead of planting grass.

Weed-Suppressing Lawn Replacements

Plant	Habit	Height	Ornamental Value	Culture
Creeping Jenny (*Lysimachia nummularia* 'Aurea')	Creeping groundcover	3" to 4" (7.6 to 10.2 cm)	Lush, rounded, golden yellow leaves and a sprawling growth habit Very fast growing Needs monitoring to make sure it stays inbounds.	Prefers moist soil and partial shade but will grow in full sun with adequate watering.
Blue wood sedge (*Carex flaccosperma*)	Clumpy grasslike perennial	6" to 12" (15.2 to 30.5 cm)	Blue-green grasslike foliage and interesting seedheads that develop in the spring	Prefers shade and well-drained soil Looks better if it is cut back in late winter
Russian arborvitae (*Microbiota decussata*)	Evergreen groundcover	8" to 12" (20.3 to 30.5 cm)	Bright green foliage turning shades of brown to purple in winter	Prefers moist, well-drained soil and sun but tolerates dry conditions
Common periwinkle (*Vinca minor*)	Evergreen groundcover	4" to 6" (10.2 to 15.2 cm)	Dark green leaves and periwinkle blue star-shaped spring flowers	Prefers moist, well-drained soil and partial shade
Two-row stonecrop (*Sedum* × 'John Creech')	Succulent groundcover	3" to 6" (7.6 to 15.2 cm)	Small, succulent, evergreen leaves; creeping habit; and attractive pink flowers in late summer and early fall	Full sun or partial shade and well-drained, even dry soils

Creative use of groundcovers can eliminate the need for a lawn.

These are just a few examples of groundcovers that could be used to replace a traditional grass lawn, but there are many more to choose from. Just remember to choose a noninvasive groundcover that is medium-to-fast growing with a dense habit and the ability to choke out weeds. Planting one or two species that are compatible is acceptable, but don't plant a menagerie of plants that are not compatible. Whatever you choose, once established, your ground-hugging plants should reduce maintenance and look good without all of the fuss that a lawn requires.

CHAPTER 6

MAINTAINING A HEALTHY GARDEN FROM THE GROUND UP

What You Need to Know about Recycling, Composting, and Soil Management

Soil biology and the health of the soil play a huge part in supporting a sustainable garden. Soil is the foundation that sustains life in the garden. If it is lacking or compromised, sustainable gardening is unrealistic and hard to achieve. Everything starts in the soil, and it is important to investigate your soil conditions first before investing any time in enhancing your garden. While it is true that certain plants will thrive in poor, depleted soils, by making efforts to understand your soil, its drainage and friability (ability to be "crumbled"), pH, organic matter content, and the microorganisms that live in soil, your chance of garden success is much greater. While improving soil health takes time, the benefits will have long-lasting rewards. It is similar to a building contractor who creates a strong foundation on a house, uses quality building materials, and does fine-quality craftsmanship—in the end, you have a better home. The hard work and investment of time and money in your landscape will pay off in the future.

SOIL IS THE FOUNDATION

There are many ways to encourage healthy, sustainable soil.

1. Create your own compost to both recycle garden debris and return nutrients to the soil through the process of decomposition.

2. Perform regular soil tests, which will tell you the type and contents of your soil. Soil test kits are relatively inexpensive and easy to use. Store-bought tests will provide basic information about your soil. More elaborate soil tests are available from many government or university-based agricultural programs.

3. Regular and proper mulching of your garden beds will also help with the long-term health of the soil while also enhancing the aesthetics of the garden. Nothing is more satisfying than a new spring planting with a fresh blanket of clean, soft mulch to finish the job. As this mulch decomposes, it slowly adds vital nutrients to your soil.

4. Fertilizing your soil—whether with homemade compost, mulch, or processed organic fertilizer—can be done to boost plant growth and vigor. Seasonal fertilization of your garden will keep plants lush and beautiful. While there are many products on the market to choose from, my best advice is to use only what you need. A current soil test will guide you in that decision.

5. Understand your soil and how it connects to the performance of your plants. Make plant choices based on the type of soil you have.

Let's take a deeper look at each of these factors and how they influence the health of your soil and in turn, your garden.

RECYCLING ORGANIC MATTER IN THE GARDEN

Recycling is critical for long-range sustainability. By recycling, we can reduce toxic wastes and environmental pollution. Recycling also conserves natural resources, protects natural ecosystems, and encourages biological diversity, all of which enhance the long-term sustainability of the environment as a whole. One of the easiest ways to recycle in the garden is to compost.

Compost: You Scratch My Back, I'll Scratch Yours

Composting is one of the easiest and best ways to feed your plants. It is environmentally friendly and natural, unlike conventional chemical fertilizers, which can damage plants if applied too heavily or at the wrong time. Compost is organic material, such as plant matter and animal manure, that has been

Organic kitchen waste is an excellent raw material to add to your compost pile.

aged (recycled) to create a finished product known as humus. Compost is a natural form of fertilizer and an important soil amendment. Incorporating it into the soil during or after planting will enhance plant growth because compost offers nutrients and improves the soil's water-holding capacity.

Compost also encourages the presence of beneficial microorganisms in the soil. These microorganisms include bacteria, fungi, worms, and other microbes that help break down organic matter and help plants absorb nutrients and water. For example, mycorrhizae are soil fungi that develop a beneficial partnership with plant roots. This partnership allows the fungi to extract carbohydrates from the plants. The fungi colonize plant roots and become extensions of the root systems, helping plants absorb nutrients and water more effectively. This colonization can very often be found in mulch or compost piles and looks like small, white threads.

It is important to point out that, although finished compost looks like soil, it is not soil. Compost is raw organic matter that's been broken down. Soil is composed of rock particles and minerals that have unique chemical components useful for growing plants. Compost can enhance soil's chemical makeup, nutritional value, and drainage capabilities; but it cannot replace soil.

The Elements of Compost

Composting and the use of organic products to enhance soil biology and improve plant growth is a very complex procedure. It is important for gardeners to know how to compost, including what raw materials to use and how to best use the finished product. Composting and the reuse of organic products to make a natural form of fertilizer require specific materials and environments to produce a tangible benefit. A healthy compost pile needs adequate oxygen and water to support the microorganisms necessary to break down the fresh organic material.

Air and Space

By providing ample room and regularly turning your compost pile, you can properly manage and aerate this material. The turning process supplies oxygen to the compost and helps the decomposition process. Turning your compost pile every few weeks, on average, is recommended.

Moisture

Moisture is an important component for a healthy composting system. Adding water to a compost pile as needed will encourage microorganism growth. Adding water is required only during dry, hot weather. The best way to determine whether water is needed is by digging into your compost pile and checking how moist it is. If it is very dry or dusty, then add water. If it is soggy, then let it dry out a bit. You want to maintain a compost pile that is evenly moist.

Heat

Another important factor in facilitating the decomposition of compost is heat. As your compost pile grows in size and begins to decay, this process naturally generates heat. Heat helps to accelerate the decay process while killing off harmful pests and weed seeds. The combination of oxygen, moisture, and heat all encourage the presence and growth of beneficial fungi and bacteria. These fungi and bacteria are microorganisms that are vital to the decomposition process needed to produce compost. Fungi are important because they break down tough debris that is not easily consumed, while bacteria decompose material that is easier to consume.

Compost Ingredients

In addition to space, water, oxygen, and heat, you also need raw organic materials to compost. There are several schools of thought on what to put in your compost pile. Some gardeners like to mix kitchen waste such as vegetables, coffee grounds, and leafy materials with grass clippings, leaves, and wood chips. This will produce compost that is appropriate for a wide variety of uses.

Another method is to separate compostable materials into different compost piles, which allows gardeners to create certain types of finished compost for specific uses. For example, green products, such as herbaceous material, grass clippings, or kitchen waste, can be separated. Herbaceous material is soft, fleshy plant growth from annuals, perennials, and other plants. Bacteria will mostly decompose green products. "Brown" products, such as wood chips and leaves, can be composted separately and will be decomposed by fungi. These brown products are an important source of carbon.

A wide variety of raw materials can be used in your compost pile.

Once composted fully, these two different types of compost can be applied to specific plants to meet their specific needs. Green compost can be used on turf, vegetable gardens, and herbaceous borders. Brown compost can be used on shrubs and trees. The benefit of separating compost is that if you mostly have specific landscape plants such as herbaceous borders and lawns, then green compost will be most beneficial to them. If you are mostly adding compost to trees and shrubs, then brown compost will be most beneficial.

If you have a little of both in your garden, which most of us do, I recommend you mix the green and brown materials together into your pile and let nature take its course. This can easily be done by alternating layers of your green and brown raw materials in your compost pile.

This well-made compost bin provides the necessary air and light needed to turn yard waste into compost.

Composting Systems

Bin composting is an efficient composting method. It is the most accepted, popular method of turning organic material into useable fertilizer. It requires gardeners to either purchase or build a bin for holding the organic materials, which are then left to decompose. The naturally cooked and finished compost is eventually harvested for use. A cylinder of wire fencing is a simple, inexpensive way of composting your yard waste if you want to make your own bin. The cylinder can be any size you desire, based on how much space and material you have on hand. Other homemade systems include using snow fencing or wooden pallets nailed or tied together. While all these options are inexpensive and offer great flexibility, they are rather unsightly and should be placed in hidden areas of the garden.

Compost bins can also be purchased ready-made and come in a variety of styles and materials. A simple holding bin, usually made out of lightweight, durable plastic, is simply a barrel that you put your organic matter to be composted in. It does not require turning, but material in this type of bin can take 6 months to 2 years to break down completely, due to lack of aeration. Some units have a ventilation system to help the aeration process, but overall these types of compost bins make it more difficult to turn and harvest usable compost.

Rotating bins or barrels, usually made from plastic or metal, are fastened on a stand of some sort, making them easy to turn regularly. This turning motion will aerate your compost, causing faster breakdown of the material inside. It is important to have enough

Prefabricated compost bins come in all shapes and sizes.

materials to fill the rotating barrel and to not add new material to an existing batch being processed. Otherwise your compost will have material in different stages of decomposition, and this will delay the finished product. These types of composting systems can be more costly and require assembly.

Binless Composting

There are also composting systems that do not require bins at all. A binless system is simply a loose pile of organic materials that can be turned regularly—or not at all. While you will spend some time keeping the heap from sprawling out far from where you want it to, it is simple and costs nothing to the resourceful gardener who is willing to put the extra work into it.

Another binless method is just to bury yard waste and organic material in the ground. Dig a hole large enough to accommodate all your waste with at least 8 inches (20.3 cm) of soil covering it. Over time the material will decompose, and the fertile soil in this area can be used at a later date. This process does require hand digging, and decomposition may take up to a year, since aeration is not an option after the composting process is compete, the area may settle and more soil or finished compost may need to be added. Gardeners should avoid planting in this area for at least that long so the composting process has time to finish.

A Recipe for Compost

Here's what you'll need:

- Compostable materials: wood chips, grass clippings, weeds and herbaceous trimmings, hedge trimmings, coffee grounds, eggshells, fruit and vegetable trimmings, animal manures (except dog and cat)
- Compost bin or pile
- Thermometer

Pile up yard and kitchen waste—the more the better—to generate heat. A pile 3 × 3 feet (0.9 × 0.9 m) or more will retain enough heat to kill any plant pathogens and weed seeds. The optimum temperature range is 130°F to 140°F (54.4°C to 60°C). Watch the temperature! Do *not* exceed 160°F (71.1°C) for an extended period because few beneficial organisms actively decompose organic material at temperatures higher than this. Regularly turn your compost pile to aerate it. Keep your compost pile "cooking"—you will need to add new raw materials regularly to feed the pile. If you have one main area where you pile debris and you are constantly adding material, your compost pile is never really done. One solution is to locate two compost bins next to each other. When one compost pile is well aged and decayed, add new raw materials to a second bin. When thoroughly cooked, "serve" the compost to your garden. Whether you decide to separate green products from brown products or mix them together, there are rules governing what to add to your compost pile and what to avoid. Check out the table of common materials to compost and ones to throw in the trash on page 155.

A special thermometer can be used to measure how hot your compost gets.

Compostable Items versus Trash

Good	Not Good
Wood chips	Bones
Leaves	Cat litter
Grass clippings	Charcoal and briquettes
Herbaceous material	Cooked food
Hedge trimmings	Dairy products
Coffee grounds	Oily or greasy foods
Eggshells	Meat
Most fruits and vegetables	Glossy paper
Animal manure	Fish scraps

FERTILIZERS, MULCH, AND COMPOST: WORKING TOGETHER

Fertilizers

There are many formulations of fertilizer on the market, designed specifically for lawns, trees and shrubs, or herbaceous plants such as vegetables, annuals, and perennials. Fertilizers are generally divided into two groups: inorganic and organic, with inorganic formulations being composed of synthetic chemicals. Inorganic fertilizers can certainly serve a valuable role in gardening, but choosing organic fertilizers first is a more sustainable approach to gardening. An organic fertilizer is derived from plant or animal sources, such as composted or processed plant parts or manure. Fertilizers can be either slow release or fast release depending on their makeup. Organic fertilizers tend to be slower release because microorganisms in the soil typically have to break down the organic fertilizer into a more useable form. Inorganic or chemical fertilizers can either be slow or fast release depending on their formulation.

While composting allows you to make your own fertilizer, there are many organic fertilizers on the market that can be purchased. Compost is cheap to make and offers long-term benefits but is a slow-release form of plant food that takes time to work. Commercial, processed organic fertilizers that you buy from the store are also typically slow release but tend to work faster than compost. These processed fertilizers may also come in specific formulations for specific crops or plant needs. These commercial fertilizers use many different naturally derived ingredients, such as blood meal, bonemeal, amino acids, seaweed extracts, fish meal, microbial inoculants (mycorrhizae), and feather meal, among many others.

While there is obviously a cost involved with using these fertilizers on a regular basis, the advantage is that they are relatively safe and effective products and can be used for specific purposes. For example, if your soil test reveals that the soil in your garden is deficient in phosphorous, you can purchase bonemeal to effectively add a natural form of this nutrient to the soil. It is imperative to follow the application instructions on the label, and it may take more than one application to see tangible results.

Much about Mulch

Another way to naturally feed your plants *and* improve soil biology over an extended period is to mulch your garden. *Mulch* is a general term that can apply to many types of products. Most mulch is applied to the surface of the ground in planting beds and around trees, shrubs, flowers, and vegetables. Often the mulches used in the garden are influenced by the area where you live and the raw materials that are available in your part of the world.

A 1- to 2-inch (2.5 to 5 cm) layer of mulch will offer a neat, manicured appearance to shrubs, trees, and other ornamentals.

Typically, the types of mulch used in a garden are wood chips, pine straw, cocoa shells, shredded leaves, or sifted compost. Mulch has many benefits, including reducing weed growth, moderating soil moisture and temperature, and enhancing the aesthetics of the garden. But the most important function of mulch is to slowly add nutrients and support the growth of beneficial microbes in the soil.

Mulch functions the same way as compost and is also quite effective as a soil amendment. The main difference is that mulch is used as a topdress rather than being incorporated into the soil, and mulch has usually not gone through the composting process yet, so it is in a more raw form. I recommend mulching at least once a year depending on how fast the material you are using decomposes. Different mulch products vary in the time it takes them to decompose, so mulching application frequency will vary.

Timing of Mulching

Mulching your landscape should be done at appropriate times during the growing season. Mulching your garden in very dry, hot weather or very wet conditions is not recommended. When mulching, soil should be evenly moist. Plants can be damaged if mulch is applied when they are stressed from heat and drought. In addition, if soil is too wet, mulching over that soil can encourage harmful diseases.

The source from which you get your mulch is crucial to the success of your garden. Adding poor-quality mulch is a waste of time and resources. There are several ways to ensure you're getting a clean, reliable source of mulch that can be used in garden beds, around vegetables, and in shrub borders. Purchase mulch from a local source or develop a relationship with a local landscape or tree care professional to supply clean mulch. Beware of processed mulch, which can be purchased from local hardware stores or retail outlets. Gardeners should avoid processed mulch that has been dyed a specific color, such as red, blue, black, and so forth, as those products contain chemicals and are *not* consistent with a truly sustainable garden strategy.

Gardeners can also purchase a small shredder that can accommodate small branches and leaves. These fairly inexpensive machines can be either electric or gas-powered. The advantage of at-home shredding is that you can recycle yard waste into mulch and reuse it very efficiently.

TOO MUCH OF A GOOD THING

Clean, shredded mulch is ideal for many garden applications.

It is important not to overdo mulch. Too much of a good thing can be quite harmful. Mulch should be applied at a depth of 1 to 2 inches (2.5 to 5.1 cm) either in the spring or the fall. Mulching at depths of 3 inches (7.6 cm) or more, or mounding mulch around the base of your plants, especially trees and shrubs, can have catastrophic results. Why? Plants need oxygen, just as we do. Too much mulch causes a low-oxygen environment around the plant's roots. This causes plant roots to grow toward the surface of the soil in search of oxygen. As these surface roots grow and become established, they can eventually grow around the base of a tree or shrub until the root cuts off the water flow and nutrient uptake of the plant. This "strangling" effect can even cause plants to weaken and fall over in windstorms. In addition, overmulched plants can develop a secondary root system close to the surface of the soil, which stresses plants even more, making them more sensitive to drought and other weather extremes. Plants in the rhododendron family (*Ericaceae*) are examples of plants that are quite sensitive to overmulching and burying roots too deep.

THE PLANT-SOIL CONNECTION

Soil health and biology is, without question, the most important factor in growing plants successfully. Plant life begins in the soil, and without a good growing medium, plants will perform poorly. Poor soil conditions can suppress plant growth, reduce flowering and fruit production, and cause plants to be more susceptible to drought, pest problems, and other environmental challenges. When creating a garden, it is essential to look at the landscape from the ground up, with soils being the first and most important aspect. It doesn't particularly matter whether your garden is in sun or shade, or is windy or protected; if the soil is of poor quality, compacted, or infertile, plants will not grow successfully. Too often we tend to overlook or underestimate the value of soil biology. But great gardeners spend a lot of time cultivating and enriching the soil, knowing that this time investment will pay long-term dividends. This type of long-range planning is very complementary and consistent with sustainable gardening.

Rich, organic soil is ideal for growing a wide variety of plant life.

Life within the Soil

Soil biology, or the life within the soil, is quite complex and is vital to growing plants successfully. Gardeners who ignore or underestimate the importance of healthy, viable soil are doomed to fail. There are several key factors that one should be aware of regarding soil. These factors include soil type, drainage, soil pH, soil horizons or layers, organic matter content, and levels of primary and secondary minerals in the soil. This sounds like a lot to consider, but it really is fairly easy to ascertain this information—and it is critical before moving forward.

Soils vary greatly in their structure, texture, chemical makeup, biological components, and other characteristics. The three main components of soil, which make up a soil's *texture*, are sand, silt, and clay. Sand is the coarsest with the largest particle size, and clay tends to be the finest with the smallest particle size. Soil *structure* consists of pore spaces and a mixture of solids (minerals), water, and air (gas). Because of their particle size, soils with clay or silt components are typically more prone to compaction than sandy soils. Compacted soil can suppress plant growth, limit root expansion, and predispose plants to disease.

Worms are excellent at turning your raw materials into usable compost.

Plant and Soil Compatibility

It is very important to group plants with like needs in the same areas of the garden. It is also important to know the basic makeup of your soil because different soils react in different ways—that is, soils all have various levels of reactivity, or cation exchange capacity (CEC). CEC refers to soil fertility and the ability of the soil to retain nutrients. Soils higher in clay or organic matter (humus) have higher CEC and, in general, tend to do a better job of retaining nutrients. There is a wide variety of plants that require a specific soil type in order to thrive. Soil can vary greatly, so knowing the most prevalent soil type in your garden is helpful.

Because soil and plants are so diverse, ensuring that your plants will thrive in the soil you have must be a priority. It is essential that you test your soil for its pH and nutrient content and determine soil texture before planting or cultivating the soil. As mentioned earlier, the most efficient way to do this is to send a soil sample to a laboratory for analysis. A full soil analysis will reveal pH, soil fertility, and the amount of organic matter in the soil. Typically, these tests also recommend what to add to the soil to address any issues that could negatively impact plant growth.

Plant Nutrition

The three main nutrients in the soil that are vital to plant growth are nitrogen (N), phosphorus (P), and potassium (K). There are also several important secondary and micronutrients that are essential to plant life, such as calcium (Ca), magnesium (Mg), and iron (Fe). It's necessary to know the levels of these nutrients in the soil at any time and how to properly add them if needed. Soil nutrient levels depend on many factors, including the type of soil, drainage, climate, and rainfall. Nutrient availability is affected by soil pH, soil type, rainfall, and other factors. Depending on the situation, these nutrients can be either deficient or abundant in the soil. Plants will often show physical signs of nutrient deficiencies such as poor growth, yellowing or chlorotic leaves, and so forth. Gardeners must keep a close eye on their plants and observe them before any problems become irreversible. Regular soil tests (once every year or two) will ensure that these types of problems do not become more challenging than they need to be.

You Don't Need a PhD to Understand Soil pH

Another important factor that determines that plants you can grow is the soil pH. The pH of the soil is typically discussed in relative terms and based on different levels or ranges of alkalinity or acidity. Soil pH is measured on a scale from 1 to 14, in which 7.0 is neutral. Soil pH below 7.0 is considered acidic; soil pH above 7.0 is alkaline. For example, a soil pH of 6.8 would be considered slightly acidic, whereas a pH of 4.5 would be considered

A soil pH test kit is essential for determining how acidic or alkaline your soil is.

very acidic. It is relatively easy to determine your soil pH by purchasing a soil pH meter or test kit from a local retail garden center or nursery.

Once you have determined the pH of the soil, you can evaluate the needs of your landscape. Certain plant groups, such as dogwood, holly, and rhododendron, *prefer* acidic soil, whereas other ornamentals, such as lilac and forsythia, are adapted to neutral or alkaline soil.

In the case of new plantings, you can measure the soil pH and select your plants based on that information. In the case of an established landscape, soil pH poses a much greater dilemma. Often gardeners must deal with existing plantings in poor health, trying to determine what the cause is. The first place to look is in the soil. If the soil pH is not adequate to grow the types of plants already in your landscape, you may have to take steps to alter the pH of the soil.

Soil pH can be raised or lowered by adding products to the soil, but this is a slow and tedious process that takes time and patience. There is no quick fix for altering soil pH levels. Garden lime (calcium carbonate) can be added to raise the pH, whereas aluminum sulfate or elemental sulfur will acidify or lower the pH of the soil. Generally speaking, nitrogen compounds in synthetic fertilizers, such as ammonium nitrate, applied in small intervals, will acidify soil over time. Natural products, such as animal manures and composted leaves, also help acidify the soil. However, soil pH should be changed gradually, over an extended period of time. Trying to change soil pH too quickly can damage plants. This is why it is so important to read and follow the directions on the label of products such as fertilizers, limestone products, and other soil amendments. In fact, with many products, plants can become damaged, stunted, or even killed by applying these products incorrectly or at toxic levels. With lime, even if plant damage does not occur, applying at the incorrect rate can waste time, product, and money. Remember that the main principle of sustainability is reducing waste. Throwing time and money away by overapplying garden products is counterproductive to that principle.

Sustainable gardeners care for their gardens from the ground up by understanding their soil conditions and making smart plant choices based on them.

Common Plant Nutrients

Primary Nutrients	Secondary Nutrients	Micronutrients
Nitrogen (N)	Calcium (Ca)	Iron (Fe)
Phosphorus (P)	Magnesium (Mg)	Manganese (Mn)
Potassium (K)	Sulfur (S)	Boron (B)
		Zinc (Zn)
		Copper (Cu)
		Chlorine (Cl)
		Molybdenum (Mo)

A SUSTAINABLE LIFESTYLE

Living a sustainable lifestyle reaches far beyond just the garden setting and can influence all aspects of our life. Sustainability affects what we eat, what we wear, how we spend our free time, what we drive, and even where we live. In addition to cultivating and maintaining a healthy garden, many consumer products today are geared toward living a more sustainable lifestyle. A significant component of sustainability is using green products. These green products include everything from recycled wood products to recycled plastic that can be used as building supplies to more energy-efficient lights and battery-operated power tools.

Sustainability offers so many opportunities for gardeners to live a greener life. Going green provides us with the products and the tools necessary to better maintain and sustain a healthy garden. Sound garden maintenance practices—proper pruning and planting, recycling, reducing waste, and paying close attention to our soils—can provide lasting benefits. With new scientific discoveries and innovative trends that reduce our needs and the consumption of our natural resources, we can all reap the benefits of an effective sustainability plan.

Start living a more sustainable life—in the garden!

GARDEN RESOURCES AND PLANT SOURCES

This list highlights a few of my favorite garden websites and plant sources that will expand your knowledge and appreciation for plants. These resources will also provide helpful information on garden supplies while also offering valuable information on environmentally friendly practices that will help you along the way to becoming a more sustainable gardener.

A Way to Garden www.awaytogarden.com

American Horticultural Society www.ahsgardening.org

American Public Gardens Association www.publicgardens.org

Arrowhead Alpines www.arrowhead-alpines.com

Brent & Becky's Bulbs www.brentandbeckysbulbs.com

Broken Arrow Nursery www.brokenarrownursery.com

Camellia Forest Nursery www.camforest.com

Collector's Nursery www.collectorsnursery.com

The Cook's Garden www.cooksgarden.com

Cornell Cooperative Extension www.cce.cornell.edu

Cutting Edge Plants www.cuttingedgeplants.com

Daniel J. Hinkley, Plantsman www.danieljhinkley.com

Direct Gardening Association http://directgardeningassociation.com

Fancy Fronds Nursery www.fancyfrondsnursery.com

The Flower Factory www.theflowerfactorynursery.com

Forestfarm at Pacifica www.forestfarm.com

Gardener's Supply Company www.gardeners.com

Gardens Alive www.gardensalive.com

Heritage Perennials www.perennials.com

Joy Creek Nursery www.joycreek.com

Missouri Botanical Garden Plant Finder
www.missouribotanicalgarden.org/plantfinder/plantfindersearch.aspx

The Nature Conservancy www.nature.org

Old House Gardens www.oldhousegardens.com

Online Plant Guide www.onlineplantguide.com

Pine Knot Farms www.pineknotfarms.com

Plant Delights Nursery, Inc. www.plantdel.com

Plant Ideas www.plantideas.com

Prairie Nursery, Inc. www.prairienursery.com

RareFind Nursery www.rarefindnursery.com

Royal Horticultural Society www.rhs.org.uk

Sandy Mush Herb Nursery www.sandymushherbs.com

Song Sparrow www.songsparrow.com

UCONN Plant Database www.hort.uconn.edu/plants

University of Georgia Trial Gardens http://ugatrial.hort.uga.edu

Woodlanders Nursery www.woodlanders.net

MEET VINCENT A. SIMEONE

VINCENT A. SIMEONE is an experienced lecturer, instructor, and horticultural consultant who has worked in the horticultural field for over three decades. He has spoken to hundreds of groups nationwide and has appeared on several garden shows, including *Martha Stewart Living* and shows on HGTV. Vincent annually presents horticultural lectures, workshops, and tours to garden clubs; plant societies; professional landscape, nursery, and arboricultural trade associations; and academic institutions. Lecture topics range from plant identification, woody plant selection and use, historic landscape preservation, and general plant maintenance and care. As if that's not enough, Vincent teaches horticulture classes at the New York Botanical Garden. Over the last 22 years, Vincent has assisted the renowned Dr. Allan Armitage, co-leading garden tours throughout the world.

Simeone received an AAS degree in ornamental horticulture from SUNY Farmingdale (Farmingdale, New York), and a BS in ornamental horticulture from the University of Georgia, Athens, Georgia. While at UGA, Vincent studied under well-known professors Dr. Michael Dirr and Dr. Allan Armitage. Vincent also obtained a master's degree in public administration from C. W. Post–Long Island University.

For nearly three decades, Simeone has served in public horticulture at Planting Fields Arboretum in New York, where he is the director. He has written and contributed to various gardening articles for magazines and newspapers, including the Long Island–based newspaper *Newsday*. In addition to this book for Cool Springs Press, Simeone is the author of five other books on plants. In 2010 he contributed to a first-ever textbook on public garden management. For all his accomplishments thus far, the Long Island Nursery and Landscape Association honored Simeone as Man of the Year in 2010. In 2015 he received the Distinguished Arborists Award from the New York State Arborists, ISA Chapter.

You may contact Vincent A. Simeone at: vasimeone@aol.com
www.vincentsimeone.com

INDEX

Italicized page numbers indicate a photo or its caption.